Multispectral Biometrics

David Zhang · Zhenhua Guo
Yazhuo Gong

Multispectral Biometrics

Systems and Applications

Springer

David Zhang
Biometrics Research Centre
The Hong Kong Polytechnic University
Hung Hom
Hong Kong SAR

Yazhuo Gong
University of Shanghai for Science and Technology
Shanghai
China

Zhenhua Guo
Shenzhen Key Laboratory of Broadband Network & Multimedia, Graduate School at Shenzhen
Tsinghua University
Shenzhen
China

ISBN 978-3-319-36985-3 ISBN 978-3-319-22485-5 (eBook)
DOI 10.1007/978-3-319-22485-5

Springer Cham Heidelberg New York Dordrecht London

Softcover reprint of the hardcover 1st edition 2015

Printed on acid-free paper

Springer International Publishing AG Switzerland is part of Springer Science+Business Media (www.springer.com)

Preface

Recently, biometrics technology has been one of the hot research topics in the IT field, because of the demands for accurate personal identification or verification to solve security problems in various applications, such as e-commence, Internet banking, access control, immigration, and law enforcement. In particular, after the 911 terrorist attacks, the interest in biometrics-based security solutions and applications has increased dramatically.

Although a lot of traditional biometrics technologies and systems such as fingerprint, face, palmprint, voice, and signature have been greatly development in the past decades, they are application dependent and still have some limitations. Multispectral biometrics technologies are emerging for high security requirement for their advantages: multispectral biometrics could offer a richer information source for feature extraction; multispectral biometrics is more robust to spoof attack since it is more difficult to be duplicated or counterfeited.

With the development of multispectral imaging techniques, it is possible to capture multispectral biometrics characteristics in real time. Recently, multispectral techniques have been used in biometrics authentication, such as multispectral face, multispectral iris, multispectral palmrpint, and multispectral fingerprint recognition, and some commercial multispectral biometrics systems have been pushed into the market already.

Our team certainly regards multispectral biometrics as a very potential research field and has worked on it since 2008. We are the first group that developed the multispectral hand dorsal technology and system. We built a large multispectral palmrpint database (PolyU multispectral Palmprint Database), which contains 6,000 samples collected from 500 different palms, and then published it online since 2010. Until now, this database has been downloaded by many researchers. This work was followed with more extensive investigations into multispectral palmprint technology, and this research has now evolved to other multispectral biometrics field. Then, a number of algorithms have been proposed for these multispectral biometrics technologies, including segmentation approaches, feature extraction methodologies, matching strategies, and classification ideas. Both this explosion of interest and this diversity of approaches have been reflected in the wide range of recently published technical papers.

This book seeks to gather and present current knowledge relevant to the basic concepts, definition, and characteristic features of multispectral biometrics technology in a unified way, and demonstrates some multispectral biometric identification system prototypes. We hope thereby to provide readers with a concrete survey of the field in one volume. Selected chapters provide in-depth guides to specific multispectral imaging methods, algorithm designs, and implementations.

This book provides a comprehensive introduction to multispectral biometrics technologies. It is suitable for different levels of readers: Those who want to learn more about multispectral biometrics technology, and those who wish to understand, participate in, and/or develop a multispectral biometrics authentication system. We have tried to keep explanations elementary without sacrificing depth of coverage or mathematical rigor. The first part of this book explains the background of multispectral biometrics. Multispectral iris recognition is introduced in Part II. Part III presents multispectral palmprint technologies. Multispectral hand dorsal recognition is developed in Part IV.

This book is a comprehensive introduction to both theoretical and practical issues in multispectral biometrics authentication. It would serve as a textbook or as a useful reference for graduate students and researchers in the fields of computer science, electrical engineering, systems science, and information technology. Researchers and practitioners in industry and R&D laboratories' working security system design, biometrics, immigration, law enforcement, control, and pattern recognition would also find much of interest in this book.

December 2014

David Zhang
Zhenhua Guo
Yazhuo Gong

Contents

Part II Multispectral Iris Recognition

Part I
Background of Multispectral Biometrics

Chapter 1
Overview

Abstract Recently, biometrics technology is one of the hot research topics in the IT field because of the demands for accurate personal identification or verification to solve security problems in various applications. This chapter gives an all-around introduction to biometrics technologies, and the new trend: multispectral biometrics.

Keywords Biometrics · Multispectral biometrics · Identification · Verification

1.1 The Need for Biometrics

Biometrics lies in the heart of today's society. There has been an ever-growing need to automatically authenticate individuals at various occasions in our modern and automated society, such as information confidentiality, homeland security, and computer security. Traditional knowledge-based or token-based personal identification or verification is so unreliable, inconvenient, and inefficient, which is incapable to meet such a fast-pacing society. Knowledge-based approaches use "something that you know" to make a personal identification, such as password and personal identity number. Token-based approaches use "something that you have" to make a personal identification, such as passport or ID card. Since those approaches are not based on any inherent attributes of an individual to make the identification, it is unable to differentiate between an authorized person and an impostor who fraudulently acquires the "token" or "knowledge" of the authorized person. This is why biometrics identification or verification system started to be more focused in the recent years.

Biometrics involves identifying an individual based on his/her physiological or behavioral characteristics. Many parts of our body and various behaviors are embedded such information for personal identification. In fact, using biometrics for person authentication is not new, which has been implemented over thousands years, numerous research efforts have been put on this subject resulting in developing various techniques related to signal acquisition, feature extraction, matching, and classification. Most importantly, various biometrics systems, including fingerprint,

D. Zhang et al., *Multispectral Biometrics*,
DOI 10.1007/978-3-319-22485-5_1

iris, hand geometry, and voice and face recognition systems have been deployed for various applications (Jain et al. 1999).

According to the report of Acuity Market Intelligence (The Future of Biometrics 2009), the market for biometrics technologies increased around 20 % each year in the past years. Figure 1.1a shows predicted total revenues of biometrics for 2009–2017. Figure 1.1b shows comparative market share by different biometrics technologies for the year 2009.

1.1.1 Biometrics System Architecture

A biometric recognition system is a pattern recognition system. During biometric recognition, biometric traits are measured and analyzed to establish a person's identity. This process involves several stages.

Enrollment
During enrollment, a user's physical or behavioral trait is captured with a camera or sensor and placed in an electronic template. This template is securely stored in a central database or a smart card issued to the user.

Recognition
During recognition, a sensor captures a biometric trait. The trait is then analyzed with an algorithm that extracts quantifiable features, such as fingerprint minutiae or face shape. A matcher takes these features and compares them to an existing template in the enrollment database.

1.1.2 Operation Mode of a Biometrics System

A biometrics system is usually operated in three modes: enrollment, identification, and verification. But some systems only have either identification or verification modes.

Enrollment—Before a user can be verified or identified by the system, he/she must be enrolled by the biometrics system. The user's biometrics data are captured, preprocessed, and feature extracted as shown in stages 1–3 of Fig. 1.2. Then, the user's template is stored in a database or file system.

Identification—This refers to the identification of a user based solely on his/her biometrics information, without any prior knowledge about the identity of a user. Sometimes, it is referred to one-to-many matching, or recognition. It will go through stages 1–3 to create an identification template. Then, the system will retrieve all the templates from the database for the feature matching. A result of success or failure is given finally. Generally, accuracy decreases as the size of the database grows.

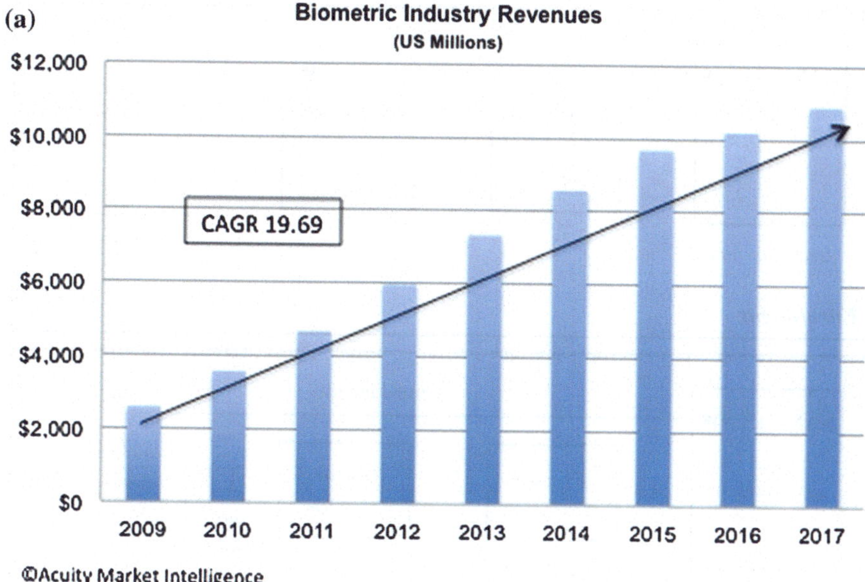

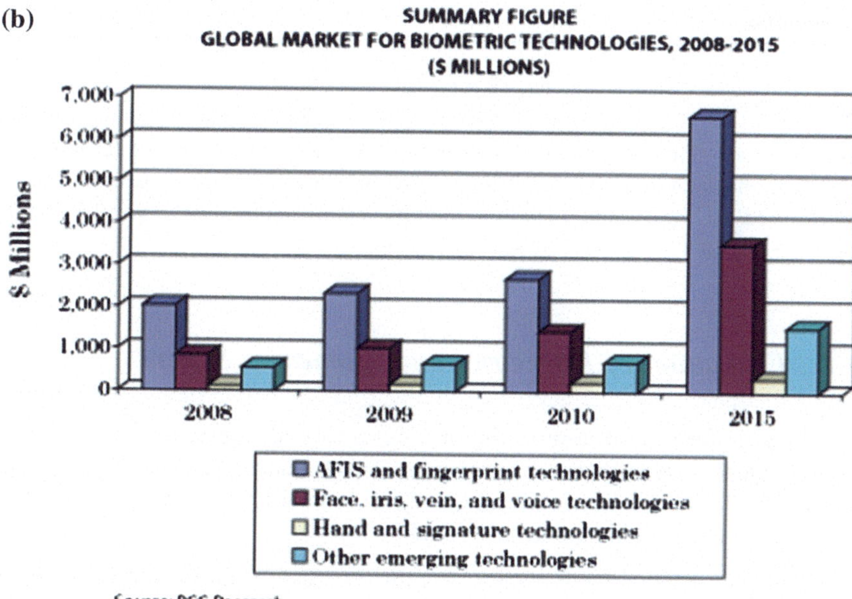

Fig. 1.1 **a** Total biometrics revenues prediction in 2009–2017 (The Future of Biometrics 2009). **b** Comparative market sharing by BCC research (Biometrics: Technologies and Global Markets 2010)

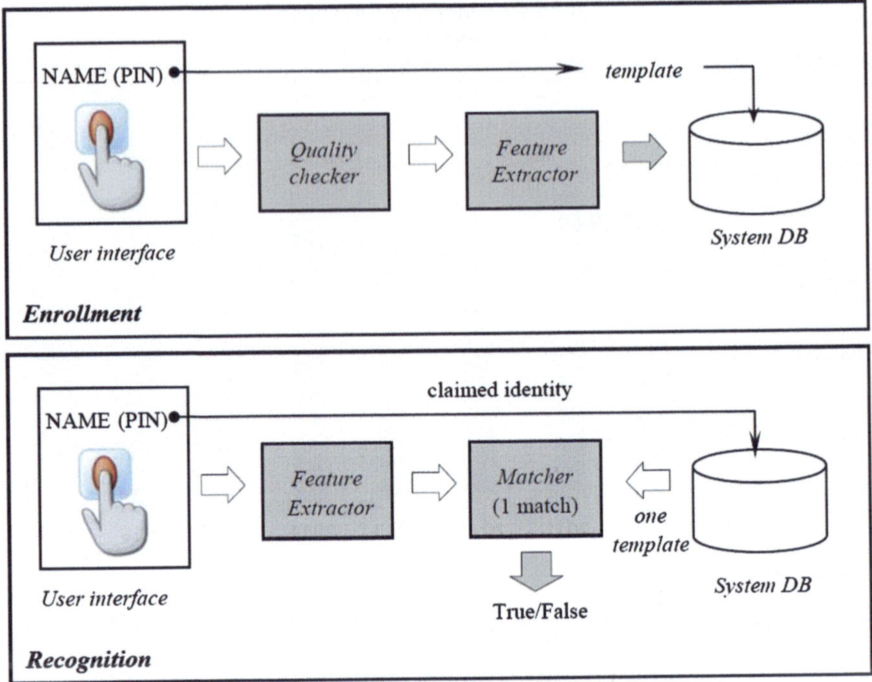

Fig. 1.2 Biometrics system architecture (What is Biometrics 2009)

Verification—This requires that an identity (ID card, smart card, or ID number) is claimed, and then, a matching of the verification template with master templates is performed to verify the person's identity claim. Sometimes, verification is referred to a 1-to-1 matching, or authentication.

1.1.3 Evaluation of Biometrics and Biometrics System

Seven factors affect the determination of a biometrics identifier, including universality, uniqueness, permanence, collectability, performance, acceptability, and circumvention. Table 1.1 summaries how three biometrics experts perceive five common biometrics technologies (Jain et al.2004).

1. Universality: Biometrics is a set of features extracted from the human body or behavior. Some human beings do not have some biometrics. For example, a worker may lose his/her fingerprint because of physical work. A dumb person does not have voice print. Universality points out the ratio of the human beings with a special biometrics.
2. Uniqueness: If a biometrics is unique, it can be used to completely distinguish any two persons in the world. The identical twins with the same genetic

Table 1.1 Perception of five common biometrics technologies by three biometrics experts (Jain et al. 2004)

	Face	Fingerprint	Hand Geometry	Iris	Palmprint
Universality	High	Medium	Medium	High	Medium
Uniqueness	Low	High	Medium	High	High
Permanence	Medium	High	Medium	High	High
Collectability	High	Medium	High	Medium	Medium
Performance	Low	High	Medium	High	High
Acceptability	High	Medium	Medium	Low	Medium
Circumvention	High	Medium	Medium	High	Medium

genotype are one of the important tests for uniqueness. Observing the similarity of a biometrics in a large database is also an important indicator for uniqueness.

3. Permanence: Many biometrics will change time by time, such as voice print, face. Iris and fingerprint, which are stable in a long period of time, are relative permanence. Permanence is described by the stability of a biometrics.
4. Collectability: Although some biometrics has high permanence, uniqueness, and universality, it cannot be used for public because of collectability. If the data collection process is too complex or requires high-cost input devices, the collectability of this biometrics is low. DNA and retina suffer from this problem.
5. Performance: The term "performance" is referred to accuracy, which is defined by two terms, (1) false acceptance rate (FAR) and (2) false rejection rate (FRR) which are controlled by a threshold. Reducing FAR (FRR) has to increase FRR (FAR). Equal error rate (EER) or crossover rate also refers accuracy.
6. Acceptability: To be a computer scientist, we should try our best to produce a user-friendly biometrics system. In fact, almost all the current biometrics systems are not physically intrusive to users, but some of them such as, retina-based recognition system, are psychologically invasive system. Retina-based recognition system requires a user to put his/her eye very close to the equipment and then infrared light passes through his/her eye in order to illuminate his/her retina for capturing an image (Miller 1994; Zhang 2000; Mexican Government 2003).
7. Circumvention: The term "circumvention" refers to how easy it is to fool the system by using an artifact or substitute.

1.2 Different Biometrics Technologies

At present, there are many different biometrics technologies. Each existing system has its own strengths and limitations. There is no perfect biometrics system until now, and the question of which one is better depends on the application. The following shows different types of biometric technologies and systems available on the market.

1.2.1 Voice Recognition Technology

Voice (speaker) recognition consists of identification and verification of the voice (Barbu 2009). Voice recognition methods encompass text-dependent and text-independent methods (Cole et al. 1997; Sammut and Squires 1995). The text-dependent methods discriminate the voice by the same utterance, such as specifically determined words, numbers, or phrases. The text-independent methods, on the other hand, recognize the voice no matter what form of words or numbers the speakers' provide. There will be waveform formed by the voice when a person speaks a word or number. The waveform is known as a voice pattern which like the fingerprints or other physical features is unique. Although everyone has a different pitch, which can be considered physical features, the human voice is classified into behavioral biometrics identifier (Barbu 2009). Figure 1.3 shows typical speaker recognition setup (Campbell 1997), and Fig. 1.4 shows a voice feature extraction scheme (Kinnuen and Li 2010). However, it is not only the qualities of the microphone and the communication channel but also the aging, medical conditions, or even emotional status can affect the behavioral part of the speech of a person (Mastali 2010). Some researchers have developed the smart home which based on the voice recognition (Baygin 2012). In the home, you can control the equipment by your voice such as turning on a lamp, close the curtain, and so on.

1.2.2 Signature Recognition Technology

Signature recognition represents an important biometrics recognition field. It has a long tradition in many common commercial fields. The technology considered to be

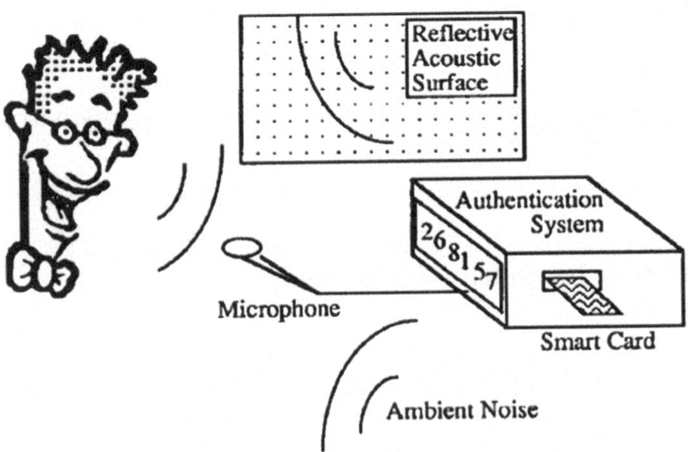

Fig. 1.3 Typical speaker recognition setup (Campbell 1997)

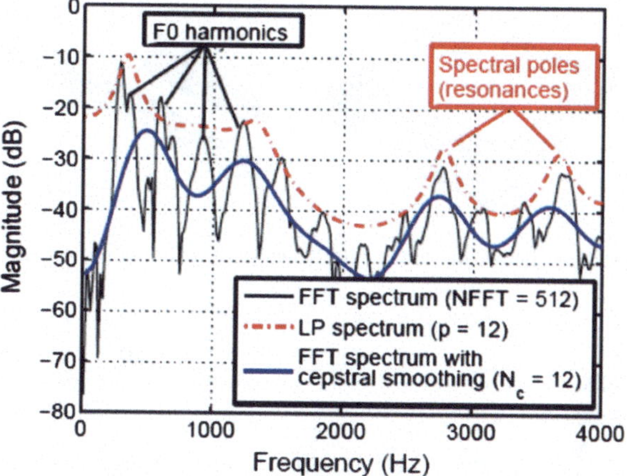

Fig. 1.4 A voice feature extraction scheme (Kinnuen and Li 2010)

a behavioral biometrics generally could be divided into two types, namely online and off-line signature recognition. The online signature recognition is mainly based on the 1-dimensional (1-D) features such as pressure, velocity, and acceleration of the signature. But the off-line signature recognition is mainly based on the static images of the signed words. Some online signature recognitions not only deal with the time domain information described above but also process the static signature images. Figure 1.5 shows a traditional signature and an online signature (Zhang et al. 2011). Figure 1.6 shows some acquisition devices (Zhang et al. 2011), and Fig. 1.7 shows typical feature extraction.

The online signature verification system developed by Prof. Berrin Yanikoglu and Alisher Kholmatov has won the first place at The First International Signature Verification Competition (SVC 2004) organized in conjunction with the First International Conference on Biometric Authentication (ICBA 2004). The recognition rates were 2.8 % EER for skilled forgery tests (Signature Verification 2002).

Fig. 1.5 A traditional signature and an online signature (Zhang et al. 2011)

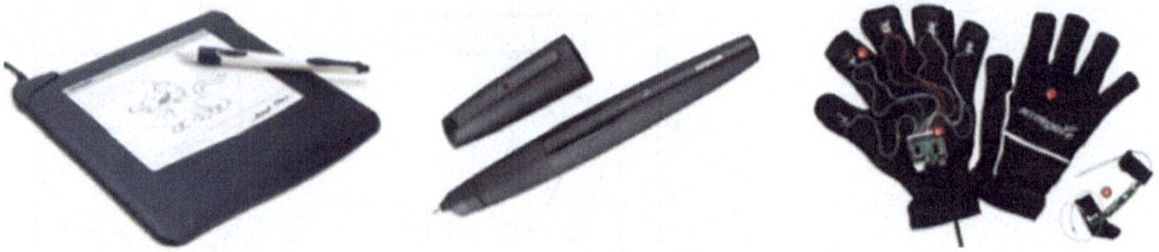

Fig. 1.6 Digitizing tablet, electronic pen, and data gloves (Zhang et al. 2011)

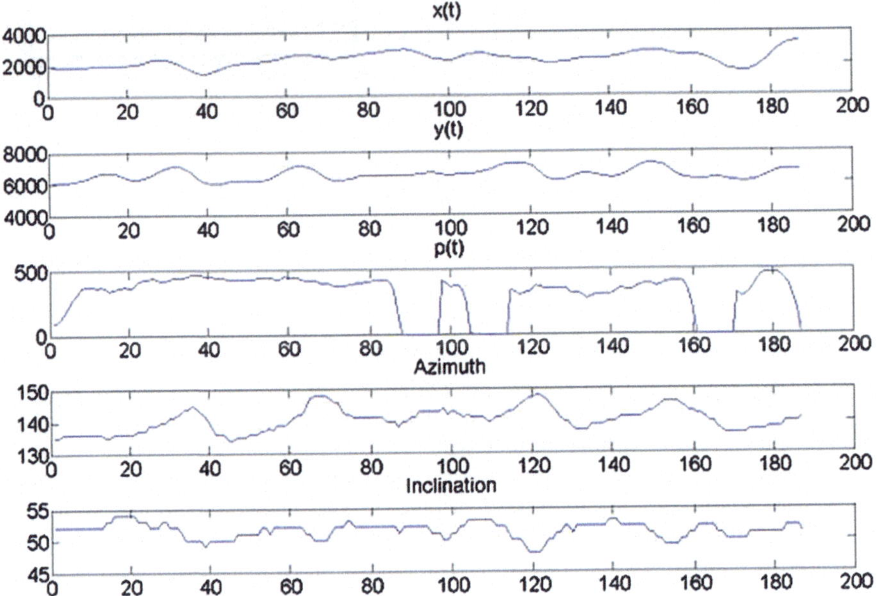

Fig. 1.7 Example of dynamic information of a signature (Signature recognition 2015)

1.2.3 Iris Recognition Technology

Iris recognition is one of the most effective biometrics technologies, being able to accurately identify the identities of more than thousand persons in real time (Pankanti et al. 2000; Zhang 2000; Jain and Pankanti 2001; Jain et al. 2004; Daugman 1993). The iris is the colored ring that surrounds the pupil. A camera using visible and infrared light scans the iris and creates a 512-byte biometrics template based on the characteristics of the iris tissue, such as rings, furrows, and freckles.

An iris recognition system such as IrisAccess™ from Iridian Technologies, Inc., provides a very high level of accuracy and security. Its scalability and fast processing power fulfills the strict requirements of today's marketplace, but it is expensive and users regard it as intrusive. It is suitable for high security areas such as nuclear plants or airplane control rooms. On the other hand, it is not appropriate in areas which require frequent authentication processes, such as logging onto a computer. Figure 1.8 shows an iris camera manufactured by LG. Figure 1.9 illustrates of the area of iris. Recently, iris on the move becomes popular. It is an approach to acquiring iris images suitable for iris recognition while minimizing the

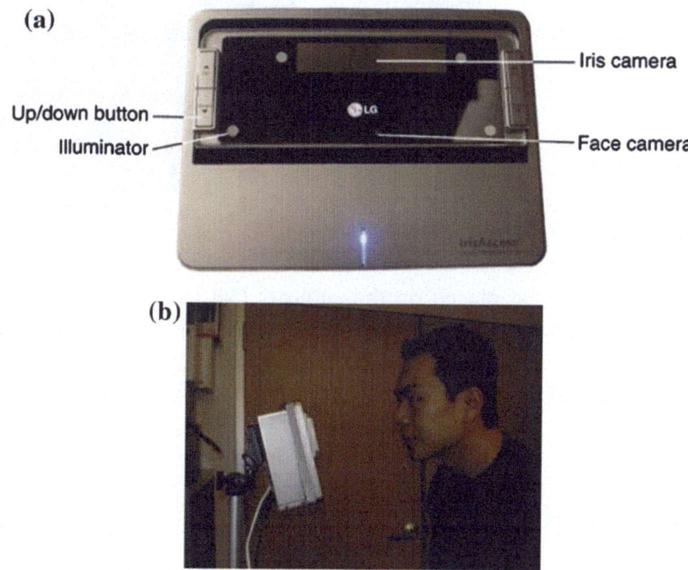

Fig. 1.8 Iris recognition overview (Li and Jain 2009). **a** LG iCAM4000 iris camera. **b** Example picture to show how to use it

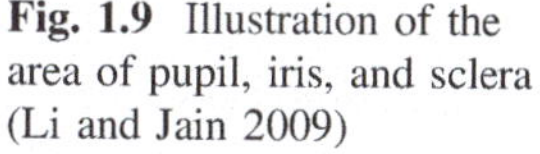

Fig. 1.9 Illustration of the area of pupil, iris, and sclera (Li and Jain 2009)

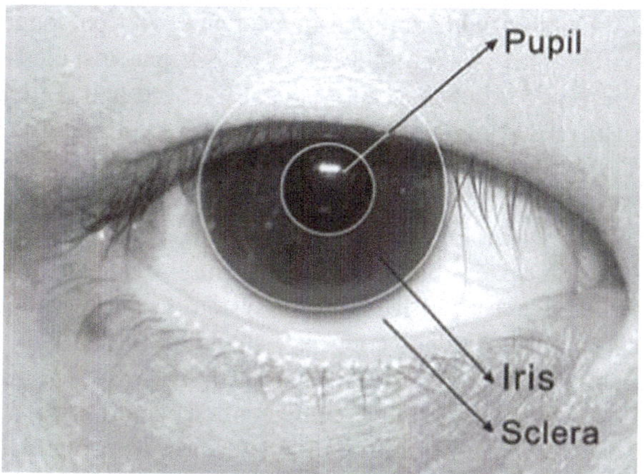

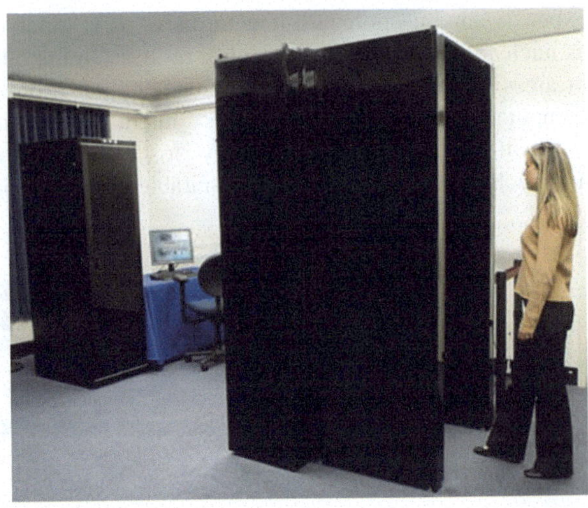

Fig. 1.10 An iris on the move portal system (Li and Jain 2009)

constraints that need to be placed on the subject. The iris on move systems have been designed to take advantage of the motion of the subject toward the camera, to avoid the need for the subjects to position themselves at the focus of the system—they will walk through it naturally as shown in Fig. 1.10.

1.2.4 Face Recognition Technology

Compared to others biometrics, face verification is low cost, needing only a camera mounted in a suitable position such as the entrance of a physical access control area. For verification purposes, it captures the physical characteristics such as the upper outlines of the eye sockets, the areas around the cheekbones, and the sides of the mouth. Face scanning is suitable in environments where screening and surveillance are required with minimal interference with passengers.

The state of Virginia in the USA has installed face recognition cameras on Virginia's beaches to automatically record and compare their faces with images of suspected criminals and runaways. However, the user acceptance of facial scanning is lower than that of fingerprints, according to an IBG Report. Figure 1.11 shows face recognition based on geometrical features.

The London Borough of Newham, in the UK, previously trialed a facial recognition system built into their borough-wide CCTV system. The German Federal Police use a facial recognition system to allow voluntary subscribers to pass fully automated border controls at Frankfurt Rhein-Main international airport. Subscribers need to be European Union or Swiss citizens. Since 2005, the German Federal Criminal Police Office offers centralized facial recognition on mugshot images for all German police agencies. The Fig. 1.12 shows a facial recognition scanner in Swiss (Facial recognition system 2015).

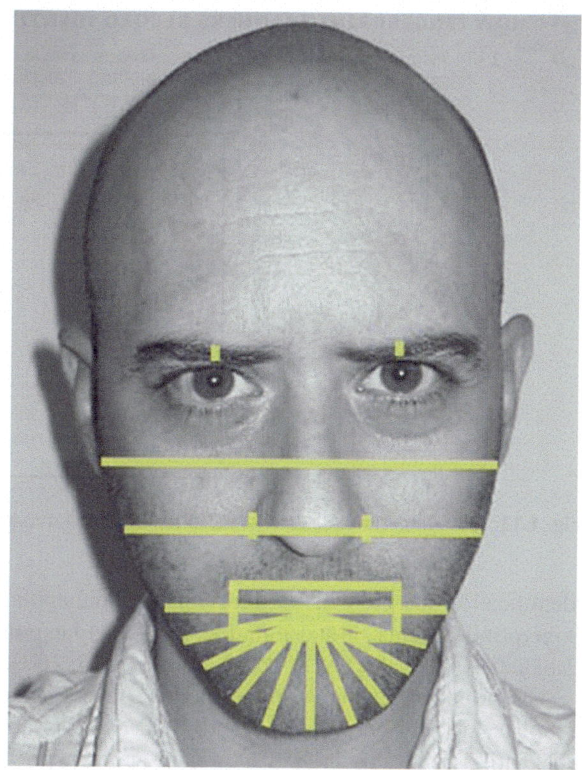

Fig. 1.11 Face recognition based on the geometrical features (Li and Jain 2009)

Fig. 1.12 Swiss European surveillance: facial recognition and vehicle make, model, color, and license plate reader (Facial recognition system 2015)

1.2.5 Fingerprint Recognition Technology

Automatic fingerprint identification began in the early 1970s. At that time, fingerprint verification had been used for law enforcement. From 1980s, the rapid development of personal computer and fingerprint scanner was started; consequently, fingerprint

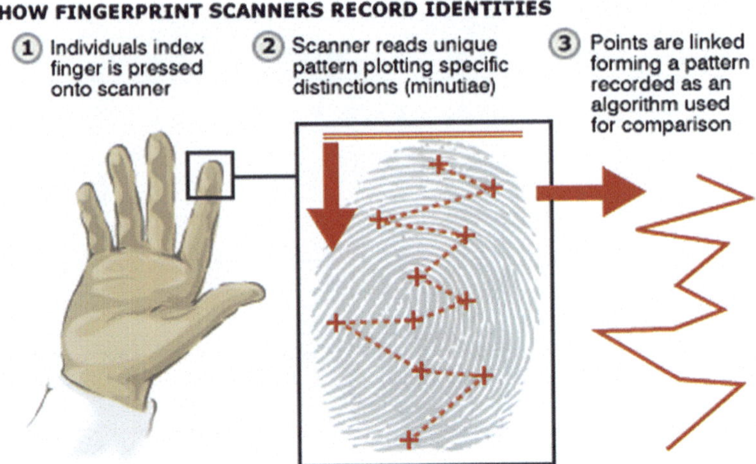

Fig. 1.13 How fingerprint scanners record identities (Motorists to give fingerprint 2006)

identification started to be used for non-criminal applications (Zhang 2000). Current fingerprint systems utilize minutiae and singular points as the features (Jain and Pankanti 2001; Jain et al. 2004; Daugman 1993, 2003; McGuire 2002).

The most promising minutiae points are extracted from an image to create a template, usually between 250 and 1000 bytes in size (Jain et al. 1997; Jain and Pankanti 2001; Ratha et al. 1996; Karu and Jain 1996; Cappelli et al. 1999; Maio and Maltoni 1997; Berry 1994). It is the most widely used biometrics technology in the world. Its small chip size, ease of acquisition, and high accuracy make it the most popular biometrics technology since the 1980s. However, some people may have fingerprints worn away due to hand work and some old people may have many small creases around their fingerprints, lowering the system's performance. In addition, the fingerprint acquisition process is sometimes associated with criminality, causing some users to feel uncomfortable with it. A typical fingerprint system is shown in Fig. 1.13.

1.2.6 Palmprint Recognition Technology

Palmprint is concerned with the inner surface of a hand and looks at line patterns and surface shape. A palm is covered with the same kind of skin as the fingertips, and it is larger than a fingertip in size. Therefore, it is quite natural to think of using palmprint to recognize a person. Because of the rich features including texture, principal lines, and wrinkles on palmprints, it is believed that they contain enough stable and distinctive information for separating an individual from a large population.

There have been some companies, including NEC and PRINTRAK, which have developed several palmprint systems for criminal applications (Jain et al. 1997; Miller 1994). On the basis of fingerprint technology, their systems exploit high-resolution palmprint images to extract the detailed features like minutiae for matching the latent prints. Such approach is not suitable for developing a palmprint authentication system for civil applications, which requires a fast, accurate, and reliable method for the personal identification. The Hong Kong Polytechnic University developed a novel palmprint authentication system to fulfill such requirements, as shown in Fig. 1.14. Figure 1.14a shows a CCD camera-based 2-D palmprint acquisition device, and Fig. 1.14b is a palmprint image collected by this device.

1.2.7 Hand Geometry Recognition Technology

Hand geometry recognition is one of the oldest biometrics technologies used for automatic person authentication. Hand geometry requires only small feature size, including the length, width, thickness, and surface area of the hand or fingers of a user, as shown in Fig. 1.15a.

There is a project called INSPASS (Immigration and Naturalization Service Passenger Accelerated Service System) which allows frequent travelers to use 3-D hand geometry at several international airports such as Los Angeles, Washington, and New York. Qualified passengers enroll in the service to receive a magnetic stripe card with their hand features encoded. Then, they can simply swipe their card, place their hand on the interface panel, and proceed to the customs gate to avoid the long airport queues. Several housing construction companies in Hong Kong have adopted the hand geometry for the employee attendance record in their construction sites, as shown in Fig. 1.15b. A smart card is used to store the hand shape information and employee details. Employees verify their identities by their hand features against the features stored in the smart card as they enter or exit the construction site. This measure supports control of access to sites and aids in wage calculations.

Hand geometry has several advantages over other biometrics, including small feature size, less invasive, more convenient, and low cost of computation as a result of using low-resolution images (Sanchez-Reillo et al. 2000; Sanchez-Reillo and Sanchez-Marcos 2000). But the current hand geometry system suffers from high cost and low accuracy (Pankanti et al. 2000). In addition, uniqueness of the hand features is not guaranteed, making it unfavorable to be used in one-to-many identification applications.

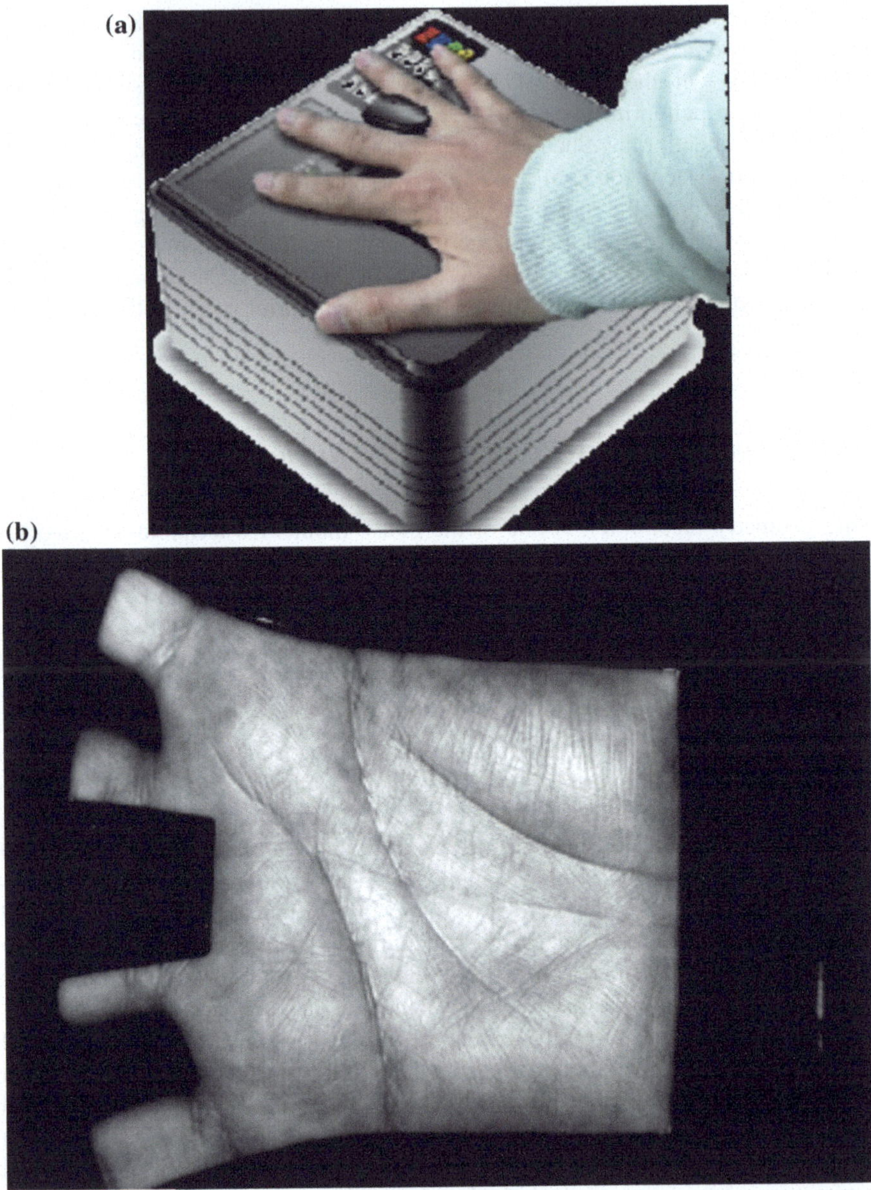

Fig. 1.14 **a** CCD camera-based 2-D palmprint acquisition device. **b** A palmprint image (*right*) that is collected by the device

(a)

(b)

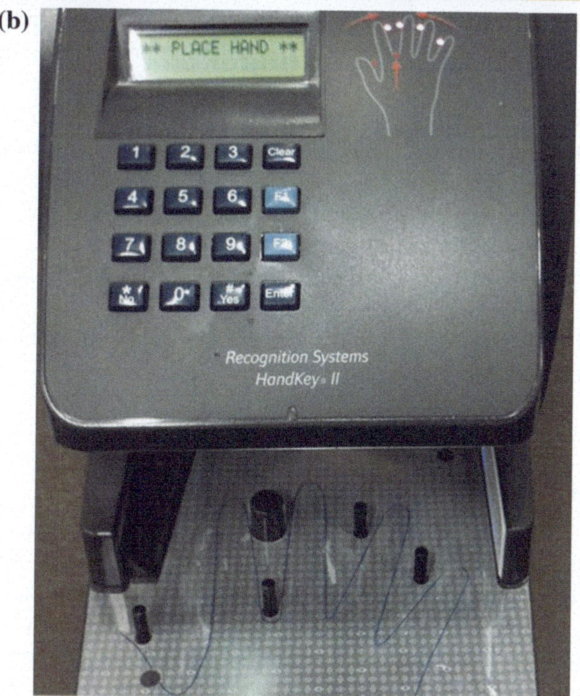

Fig. 1.15 **a** The features of a hand geometry system. **b** A hand geometry reading system (Hand Geometry 2014)

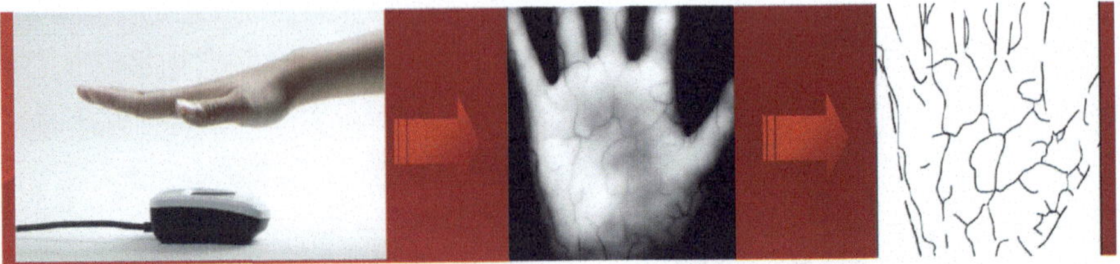

Fig. 1.16 The process of capturing of image of the hand vein patterns (Fujitsu Laboratories Limited 2014)

1.2.8 Palm Vein Recognition Technology

Fujitsu Laboratories Limited developed a new type of biometrics authentication system (PalmSecure™) which verifies a person's identity by the pattern of the veins in his/her palms (Fujitsu Laboratories Limited 2002). The PalmSecure™ works by capturing a person's vein pattern image while radiating it with near-infrared rays (Fujitsu Laboratories Limited 2011a, b). The scanning process is extremely fast and does not involve any contact between the sensor and the person being scanned. The process is shown in the Fig. 1.16 (Fujitsu Laboratories Limited 2011a, b).

Fujitsu piloted the palm vein recognition-based solution from January to June of 2011 at Boca Ciego High School. The result was tremendous: a 98 % first-scan transaction success rate with wait times in their busy lunch lines dropping from 15 min to a paltry 7 min. Starting in the fall of the 2011–2012 school year, all Pinellas County Schools will rely on the Fujitsu PalmSecure biometric solution to handle cafeteria transaction for the tens of thousands of students, across all of their 46 middle and high schools, who take part in their daily snack and lunch service program (Fujitsu Limited 2014).

1.3 A New Trend: Multispectral Biometrics

Versatility, usability, and security are some of the required characteristics of biometric system. Such system must have the capability to acquire and process biometric data at different times of day and night, in a variety of weather and environmental conditions, and be resistant to spoofing attempts. Multispectral biometrics is one of the few technologies shown to solve many of the aforementioned issues (Li and Jain 2009).

Multispectral imaging is possible to simultaneously capture images of an object in the visible spectrum and beyond. It has been extensively used in the field of remote sensing, medical imaging, and compute vision to analyze information in multiple bands of the electromagnetic spectrum. Figure 1.17 shows two images of a person wearing sunglasses, one in the visible light and the other in the near infrared (NIR). As shown in Fig. 1.17, multispectral imaging could provide more information

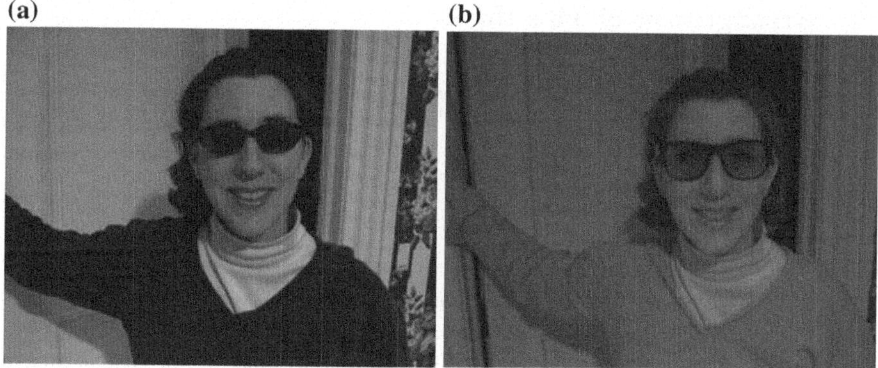

Fig. 1.17 Visible and NIR images of a subject wearing sun glasses. Images are acquired using an IR-Enabled Sony DSC-S30 Digital Camera & X-Nite780 nm filter under incandescent light (Infrared Filter 2014)

than single modality. Usually, complementary features could be extracted. Thus, better recognition and spoof detection ability are easy to be achieved.

Wavelengths covering the visible spectrum all the way to the long-wave infrared (LWIR) have been used in the analysis of biometric data. Figure 1.18 displays the distribution and values of wavelengths along the electromagnetic spectrum, with emphasis on the most used bands in biometrics. The visible spectrum comprises wavelengths between 400 and 750 nm. Bands of lengths from 750 to 1400 nm represent the NIR, while wavelengths between 1.4 and 3 μm are called short-wave infrared (SWIR). Bands from 3 to 8 μm are said to be in the mid-wave infrared (MWIR) and bands between 8 and 15 μm are called LWIR. Mid wave and long wave are also referred to as thermal infrared (Li and Jain 2009).

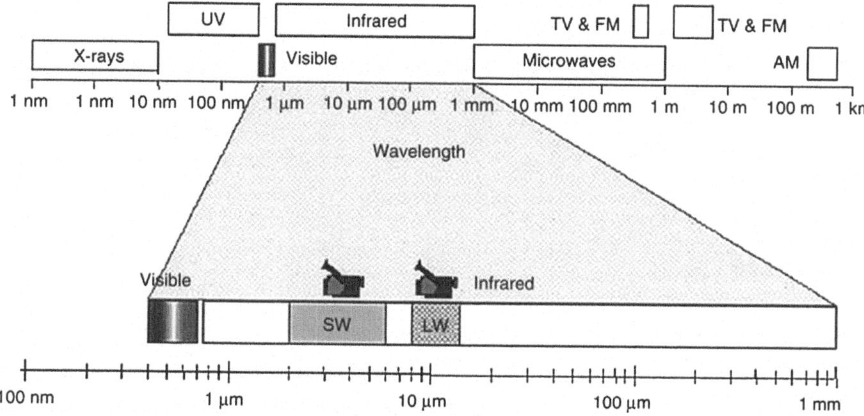

Fig. 1.18 Bands of the electromagnetic spectrum and their wavelengths (Li and Jain 2009)

1.4 Arrangement of This Book

In this work, we would like to summarize our multispectral biometrics work. Its twelve chapters are in five parts, covering multispectral biometrics technology from the hardware design of multispectral imaging systems, spectral feature selection, feature extraction, and matching.

PART I
This chapter introduces recent developments in biometrics technologies, some key concepts in biometrics, and the new trend: multispectral biometrics technologies. Chapter 2 focuses on the main multispectral imaging technologies and its application in multispectral biometrics, and then, we introduce some typical multispectral face, multispectral iris, multispectral fingerprint, multispectral palmprint, and multispectral dorsal hand data acquisition methods, feature extraction, and matching algorithms.

PART II
This part discusses multispectral iris recognition and has three chapters. Chapter 3 introduces a multispectral iris acquisition system. Chapter 4 studies feature band selection for multispectral iris recognition. In Chap. 5, we propose a prototype of multispectral iris recognition system.

PART III
This part has three chapters. Chapter 6 introduces a multispectral palmprint acquisition system. Chapter 7 studies several algorithms' accuracy by different light sources. In Chap. 8, we further explore feature band selection for multispectral palmprint recognition.

PART IV
This part includes three chapters. Chapter 9 introduces a multispectral dorsal hand recognition system with its optimal feature band selection. Chapter 10 studies feature band selection of multispectral dorsal hand biometrics. In Chap. 11, we study and compare two biometrics, palm, and dorsal hand recognition.

PART V At the end of this book, a brief book review and future work are presented in Chap. 12.

References

Barbu T (2009) Comparing various voice recognition techniques. In: IEEE proceedings of the 5-th conference on speech technology and human-computer dialogue, pp 1–6 (doi:10.1109/SPED.2009.5156172)

Baygin M (2012) Real time voice recognition based smart home application. In: IEEE 20th signal processing and communications applications conference (SIU), pp 1–4

Berry J (1994) The history and development of fingerprinting. In: Lee HC, Gaensslen RE (eds) Advances in fingerprint technology. CRC Press, Florida, pp 1–39

Biometrics: Technologies and Global Markets (2010) http://www.bccresearch.com/market-research/information-technology/biometrics-technologies-markets-ift042c.html. Accessed 30 November 2014

Campbell J (1997) Speaker recognition: a tutorial. Proc IEEE 85:1437–1462

Cappelli R, Lumini A, Maio D, Maltoni D (1999) Fingerprint classification by directional image partitioning. IEEE Trans Pattern Anal Mach Intell 21:402–421. doi:10.1109/34.765653

Cole RA, Mariani J, Uszkoret H, Zaenen A, Zue A (1997) Survey of the state of the art in human language technology. Cambridge University Press, Cambridge

Daugman JG (1993) High confidence visual recognition of persons by a test of statistical independence. IEEE Trans Pattern Anal Mach Intell 15:1148–1161. doi:10.1109/34.244676

Daugman JG (2003) The importance of being random: statistical principles of iris recognition. Pattern Recogn 36(2):279–291

Facial recognition system (2015) http://en.wikipedia.org/wiki/Facial_recognition_system. Accessed 25 February 2015

Fujitsu Laboratories Limited (2002) Biometric mouse with palm vein pattern recognition technology. http://pr.fujitsu.com/en/news/2002/08/28.html. Accessed 4 July 2003

Fujitsu Laboratories Limited (2011a) Fujitsu PalmSecure selected as a "world changing idea" for 2011 by Scientific American Magazine. http://www.fujitsu.com/us/services/biometrics/palm-vein/SA_news.html. Accessed 18 July 2012

Fujitsu Laboratories Limited (2011b) PalmSecure—Fujitsu's world-leading authentication technology. http://www.fujitsu.com/emea/products/biometr-ics/intro.html. Accessed 18 July 2012

Fujitsu Limited (2014) Contactless palm vein authentication. http://www.fujitsu.com/jp/group/frontech/en/solutions/business-technology/security/palmsecure/sensor/. Accessed 30 Nov 2014

Hand Geometry (2014) http://en.wikipedia.org/wiki/Hand_geometry. Accessed 30 Nov 2014

ICBA (2004) http://www4.comp.polyu.edu.hk/~icba/icba2004/index.htm

Infrared Filter (2014) http://www.maxmax.com/aXRayIRExamplesSecurity.htm. Accessed 30 Nov 2014

Jain A, Pankanti S (2001) Automated fingerprint identification and imaging systems. In: Lee HC, Gaensslen RE (eds) Advances in fingerprint technology, 2nd ed. CRC Press, New York

Jain AK, Hong L, Bolle R (1997) On-line fingerprint verification. IEEE Trans Pattern Anal Mach Intell 19:302–314. doi:10.1109/34.587996

Jain AK, Bolle RM, Pankanti S (1999) Biometrics: personal identification in networked society. Hardbound, Boston

Jain AK, Ross A, Prabhakar S (2004) An introduction to biometric recognition. IEEE Trans Circ Syst Video Technol Spec Issue Image Video-Based Biometrics 14:4–20. doi:10.1109/TCSVT.2003.818349

Karu K, Jain AK (1996) Fingerprint classification. Pattern Recogn 29:389–404. doi:10.1016/0031-3203(95)00106-9

Kinnunen T, Li H (2010) An overview of text-independent speaker recognition: from features to supervectors. Speech Commun 52:12–40. doi:10.1016/j.specom.2009.08.009

Li SZ, Jain AK (2009) Encyclopedia of biometrics. Springer, US

Maio D, Maltoni D (1997) Direct gray-scale minutiae detection in fingerprints. IEEE Trans Pattern Anal Mach Intell 19:27–40. doi:10.1109/34.566808

Mastali N (2010) Authentication of subjects and devices using biometrics and identity management systems for persuasive mobile computing: a survey paper. In: IEEE fifth international conference on broadband and biomedical communications, pp 1–6 (doi:10.1109/IB2COM.2010.5723618)

McGuire D (2002) Virginia beach installs face-recognition cameras. The Washington Post. http://www.washingtonpost.com/ac2/wp-dyn/A19946-2002Jul3. Accessed 14 May 2003

Mexican Government (2003) Face recognition technology to eliminate duplicate voter registrations in upcoming presidential elections. http://www.shareholder.com/identix/ReleaseDetail.cfm?ReleaseID=53264. Accessed 15 May 2003

Miller B (1994) Vital signs of identity. IEEE Spectr 31:22–30. doi:10.1109/6.259484
Mexican Government (2003) Face recognition technology to eliminate duplicate voter registrations in upcoming presidential elections. http://www.shareholder.com/identix/ReleaseDetail.cfm?ReleaseID=53264. Accessed 15 May 2003
Motorists to give fingerpint (2006) http://news.bbc.co.uk/2/hi/uk/6170070.stm#text. Accessed 30 Nov 2014
Pankanti S, Bolle RM, Jain A (2000) Biometrics: the future of identification. IEEE Comput 33:46–49. doi:10.1109/2.820038
Ratha N, Karu K, Chen S, Jain AK (1996) A real-time matching system for large fingerprint databases. IEEE Trans Pattern Anal Mach Intell 18:799–813. doi:10.1109/34.531800
Sammut C, Squires B (1995) Automatic speaker recognition: an application of machine learning. In: Proceeding of the 12th international conference on machine learning
Sanchez-Reillo R, Sanchez-Marcos A (2000) Access control system with hand geometry verification and smart cards. IEEE Aerosp Electron Syst Mag 15:45–48. doi:10.1109/62.825671
Sanchez-Reillo R, Sanchez-Avilla C, Gonzalez-Marcos A (2000) Biometric identification through hand geometry measurements. IEEE Trans Pattern Anal Mach Intell 22:1168–1171. doi:10.1109/34.879796
Signature recognition (2015) http://en.wikipedia.org/wiki/Signature_recognition. Accessed 25, 2015
Signature Verification (2002) http://biometrics.sabanciuniv.edu/signature.html. Accessed 18 July 2012
SVC (2004) http://www.cse.ust.hk/svc2004/
The Future of Biometrics (2009) http://www.acuity-mi.com/FOB_Report.php. Accessed 30 Nov 2014
What is Biometrics (2009) http://biometrics.cse.msu.edu/info/index.html. Accessed 30 Nov 2014
Zhang D (2000) Automated biometrics: technologies and systems. Hardbound, Boston
Zhang X, Wang K, Wang Y (2011) A survey of on-line signature verification. In: Proceeding of Chinese conference on biometric recognition, pp 141–149

Chapter 2
Multispectral Biometrics Systems

Abstract Until now, many multispectral biometrics technologies and systems have been proposed. Different multispectral biometrics systems have their own characteristics. This chapter gives an overall review of multispectral imaging (MSI) techniques and their applications in biometrics.

Keywords Multispectral biometrics · Face · Fingerprint · Palmprint · Iris · Dorsal hand

2.1 Introduction

Multispectral biometrics is based on data consisting of four to ten separate images of the same biometrics trait and representing sensors' responses in different wavelengths of the electromagnetic spectrum. In contrast to conventional images, which generally represent an integrated response of a single sensor over a wide range of bands in the same spectral zone, multispectral data usually refer to multiple separate sensor(s) responses in relatively narrow spectral bands. The word multispectral was first used in space-based imaging to denote data acquired in the visible and infrared spectra. In biometrics, the word multispectral has been used to describe responses in multiple narrow bands either all in the visible, or all in the infrared, or a mixture of both. Even though the words hyperspectral and multispectral have often been used interchangeably, hyperspectral imaging usually refers to cases where the number of bands is higher than 10 and when these bands encompass more than one region of the electromagnetic spectrum, such as the visible and the infrared. Figure 2.1 shows difference between hyperspectral and multispectral.

There are mainly two kinds of methods, touch-based and touchless, to acquire multispectral biometrics data. The former method utilizes light with different wavelengths to illuminate the object and capture multiple images by a camera.

D. Zhang et al., *Multispectral Biometrics*,
DOI 10.1007/978-3-319-22485-5_2

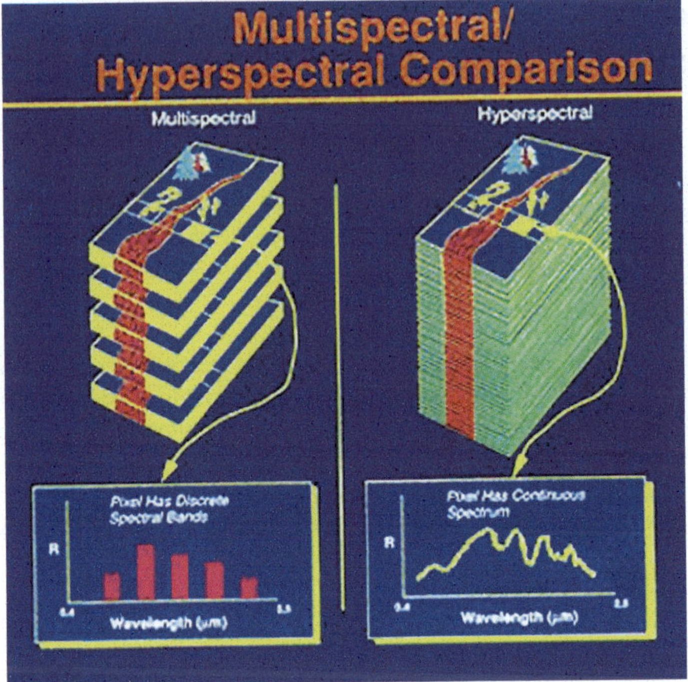

Fig. 2.1 Hyperspectral and multispectral differences (Hyperspectral imaging 2014)

It requires the subject to touch a glass or prism and is suitable to collect palmprint and fingerprint data. Figure 2.2 shows an example of multispectral hand imaging system. The latter method usually utilizes a full spectral light, such as halogen lamp, to illuminate the object. A liquid crystal tunable filter (LCTF) (shown in Fig. 2.3) or a filter wheel (shown in Fig. 2.4) with multiple filters is mounted in front of the camera. This technique is suitable to collect iris, face, palmprint, and dorsal hand data. Figure 2.5 shows an example of multispectral/hyperspectral face imaging system.

2.2 Different Biometrics Technologies

At present, there are many different multispectral biometrics technologies and systems. Each existing system has its own strengths and limitations. The following shows different multispectral biometrics applications.

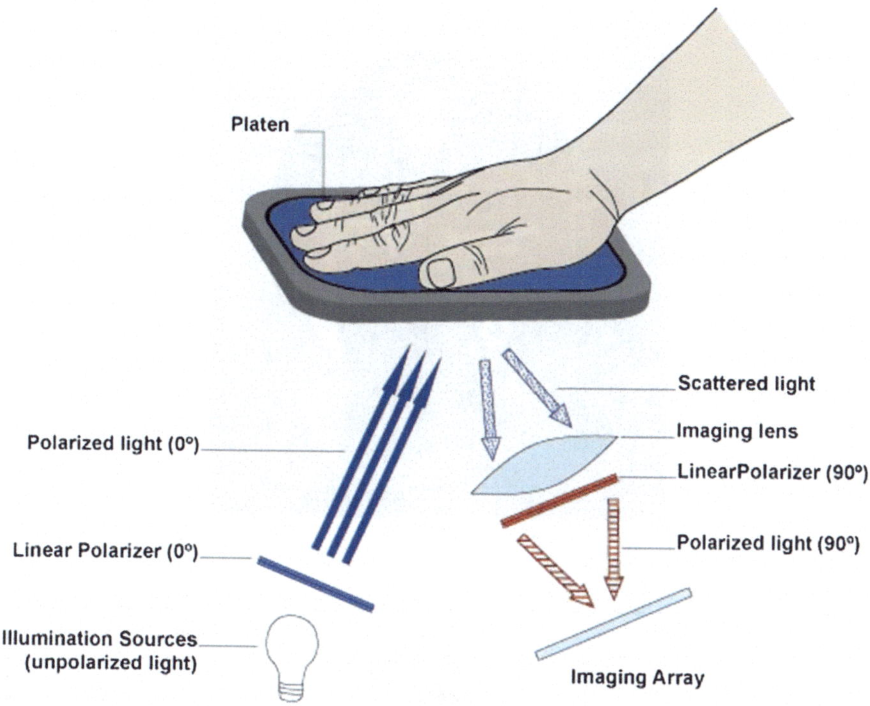

Fig. 2.2 Major optical components and layout of the multispectral whole-hand imaging system (Rowe et al. 2007)

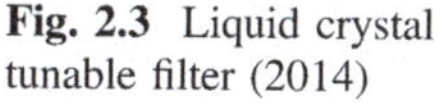
Fig. 2.3 Liquid crystal tunable filter (2014)

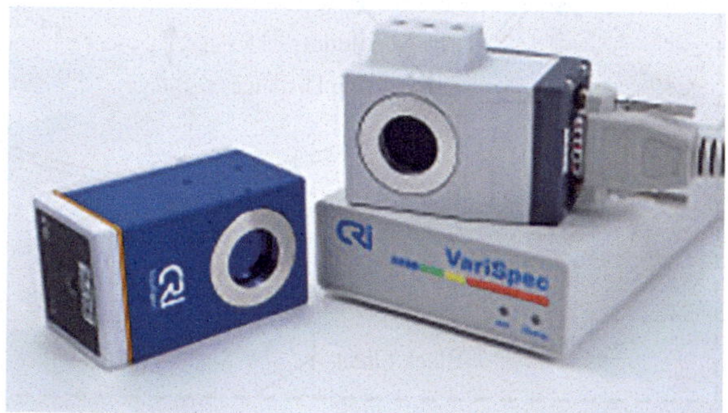

2.2.1 Multispectral Iris

Traditionally, only a narrow band of the near-infrared (NIR) spectrum (750–850 nm) was utilized for iris recognition systems since this will alleviate any physical discomfort from illumination, reduce specular reflections, and increase the amount of iris texture information captured for some of the iris colors. Commercial

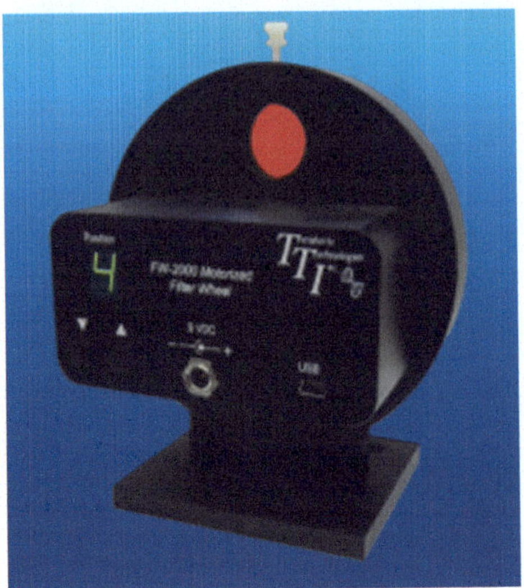

Fig. 2.4 Filter wheel (2014)

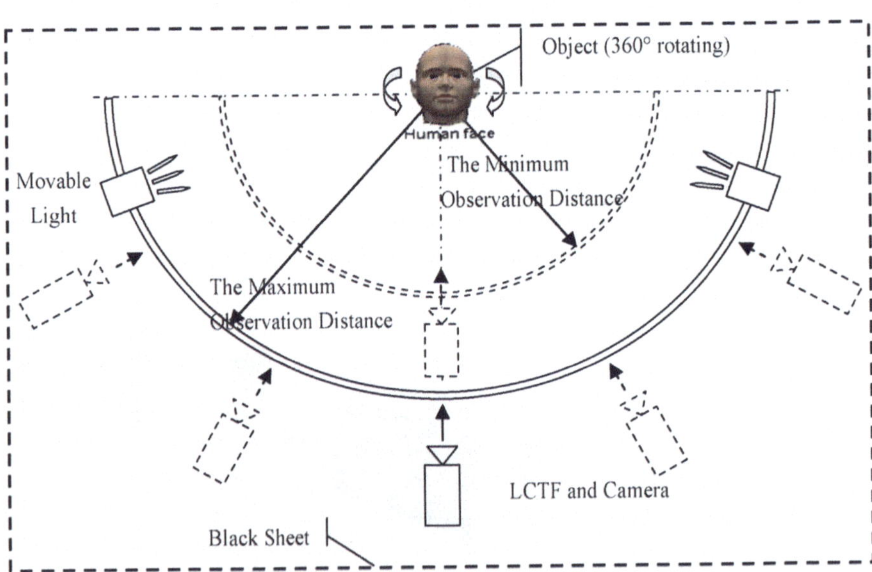

Fig. 2.5 Multispectral/hyperspectral face imaging system (Di et al. 2010)

iris recognition systems predominately operate in the NIR range of the electromagnetic spectrum. The spectrums indicate that current systems are using wavelengths that peak around 850 nm (Panasonic and Oki), with a narrow band pass. However, some systems traverse into the range of 750 nm (LG) and use multiple wavelength illumination to image the iris. The infrared light is invisible to the human eye, and the intricate textural pattern represented in different colored irides is revealed under an NIR range of illumination. The texture of the iris in IR illumination has been traditionally used as a biometrics indicator (Boyce 2006).

However, the textural content of the iris has complex components, including numerous structures and various pigments, both fibrous and cellular, which are contained on the anterior surface, including ligaments, crypts, furrows, collarettes, moles, and freckles. The NIR wavelengths can penetrate melanin, showing a texture which cannot be easily observed in the visible spectrum, but the cost is substantially high. Most of the texture presented in the NIR spectrum is only generated by the iris structures, not by the pigments. The effect of melanin, the major color-inducing compound, is negligible on the NIR wavelengths in iris recognition. But, melanin is always imaged in certain wavelength for extraction and classification, such as in tongue image processing (Liu et al. 2007).

The above study has inspired us to consider that the iris textures generated outside the NIR spectrum may have more information over those that are only generated in the NIR spectrum, because melanin can be present in the shorter wavelengths and becomes another major source of iris texture.

Previous research has shown that matching performance is not invariant to iris color and can be improved by imaging outside the NIR spectrum and that the physiological properties of the iris (e.g., the amount and distribution of melanin) impact the transmission, absorbance, and reflectance of different portions of the electromagnetic spectrum and the ability to image well-defined iris textures (Wilkerson et al. 1996). Performing multispectral fusion at the score level is proven feasible (Ross et al. 2006a, b), and the multispectral information is used to determine the authenticity of the imaged iris (Park and Kang 2007).

When we are ready to accept a fact that multispectral iris fusion can improve matching performance, which has been proven by several previous experiments, two questions need to be answered as the cornerstones that support the above conclusion. Question 1: How do the colors of the irides influence the matching performance of multispectral iris recognition? Question 2: How does the iris texture generated from the structures and pigments change with the illumination of different spectral wavelengths?

Some researchers have tried to answer Question 1 and achieved some initial results. Burge and Monaco (2009) demonstrated that iris texture increases with the frequency of illumination for lighter-colored sections of the iris and decreases for darker sections. This means that the effects of an illumination wavelength on various colored sections of the iris are not the same; sometimes the texture increases, and sometimes it decreases, depending on the color of the iris. Hence, the feasibility of multispectral iris recognition cannot be explored from only the perspective of an electromagnetic spectrum, and the colors of the irides should be studied as a very important factor, combined with illumination wavelengths.

Although these previous studies did not really answer Question 1, at least they clarified a basic principle: The accuracy of conclusions is based on the basis that the iris images used in a multispectral study should belong to a certain color classification. This is because the same research methods may produce an entirely different conclusion on the iris images of different colors.

In terms of Question 2, there is no related in-depth research and no published results. From the previous studies, we are only able to observe the phenomenon in which iris images captured across multispectral wavelengths show differences in the amount and distribution of texture and should be used for feature fusion in order to increase the diversity of iris textures, but we do not know the specific mechanisms of the iris texture that change with multispectral wavelengths. Studying the above mechanism has great significance, especially in the choice of which band of the electromagnetic spectrum can be used for iris fusion.

2.2.2 *Multispectral Fingerprint*

The multispectral imaging (MSI) technology has been widely used in the fingerprint. Testing performed to date has shown strong advantages of the MSI technology over conventional imaging methods under a variety of circumstances.

The source of the MSI fingerprint advantage is threefold. First, there are multiple anatomical features below the surface of the skin that have the same pattern as the surface fingerprint and can be imaged by MSI. This means that additional subsurface sources of signal are present for an MSI sensor to gather and compensate for poor quality or missing surface features. Second, the MSI sensor was designed to be able to collect usable biometrics data under a broad range of conditions including skin dryness, topical contaminants, poor contact between the finger and sensor, water on the finger platen, and bright ambient lighting. This sensor characteristic enhances the reliability of the MSI sensor and reduces the time and effort required by the authorized user to successfully conduct a biometrics transaction. Third, because the MSI sensor does not just measure the fingerprint but instead measures the physiological matrix in which the fingerprint exists, the resulting data provide clear indications of whether the fingerprint is taken from a living finger or some other material (Rowe et al. 2008).

For example, in the case of maritime environment, a fingerprint reader equipped with high-end MSI technology is used, which adds more to the image quality and the robustness of the data acquisition process (Fakourfar and Belongie 2009). In the case of spoof detection, the MSI fingerprint is configured to image both surface and subsurface characteristics of the finger under a variety of optical conditions. The combination of surface and subsurface imaging ensures the liveness of an object.

Nowadays, most of the MSI fingerprint image is compatible with images collected using other imaging technologies. Thus, the MSI sensor is usually incorporated into systems with other sensors, which is called multispectral multibiometrics sensing system. Commonly, the system consists of an MSI and a conventional optical sensor based on total internal reflectance (TIR). The two sensors are combined in a way that both sensors could collect data on a finger placed on the platen in approximately one second (Rowe et al. 2005). Some systems are constructed and used to collect data in a multiday, multiperson study. The sensor is based on multispectral technology that is able to provide hand shape, fingerprints, and palmprint modalities of a user's hand by a single user interaction with the sensor. Thus, it will reduce the overall size and complexity of the multibiometric system when compared to other systems that use multiple sensors, one per trait. One minor disadvantage is an increase in computational requirements due to multispectral processing of data (Rowe et al. 2007).

Along with those advantages, the MSI fingerprint has some shortcomings, such as noise and compatibility. Some have proposed a method using the texture of fingerprint images to reduce multispectral noise, which has been proved to be efficient (Khalil et al. 2009). Besides, a lot of methods have been proposed and achieved great progress, both in device and in algorithm. Still, there is a lot of work need to be done.

2.2.3 Multispectral Face

MSI is widely used technique for face recognition. Some studies regard multi-spectral face images as a kind of multimodal biometrics, thus different feature or decision level fusion schemes are explored. Zheng and Elmagbraby (2011) explore and compare four face recognition methods and their performance with multi-spectral face images and further investigate the performance improvement using multimodal score fusion. Nicolo and Natalia (2011) introduce a robust method to match visible face images against images from short-wave infrared (SWIR) spectrum. Later, Boothapati and Natalia (2013) propose a methodology for cross-matching color face images and SWIR face images reliably and accurately. Zheng et al. (2012) and Zheng (2011) propose a wavelet-based face recognition method under the framework of Gabor wavelet transform (GWT) and Hamming distance (HD) , which results in two algorithms, face pattern word (FPW) and face pattern byte (FPB). Bourlai and Bojan (2012) study the problems of intra-spectral and cross-spectral face recognition in homogeneous and heterogeneous environments and investigate the advantages and limitations of matching between different spectral bands. They also utilize both commercial and academic face matchers and performed a set of experiments indicating that the cross-photometric score-level fusion rule can improve SWIR cross-spectral matching performance. Bendada and Moulay (2010) introduce the use of local binary patterns (LBP) like texture descriptors, including LBP, local ternary patterns (LTP), and a simple differential LTP descriptor (DLT), for efficient multispectral face recognition, which is less sensitive to noise, illumination change, and facial expressions. Similarly, Akhloufi and Abdelhakim (2010) introduce a new locally adaptive texture feature descriptor called local adaptive ternary pattern (LATP) for efficient multispectral face recognition. Singh et al. (2008a, b) develop a novel formulation of multiclass support vector machine called multiclass mv-granular soft support vector machine, which uses soft labels to address the issues due to noisy and incorrectly labeled data and granular computing to make it adaptable to data distributions both globally and locally. In a multispectral face recognition application, the proposed multiclass classifier is used for dynamic selection of four options: visible spectrum face recognition, short-wave infrared face recognition, multispectral face image fusion, and multispectral match score fusion.

Some works focus on image/feature-level fusion of multispectral images. Buddharaju and Pavlidis (2007) have outlined a novel multispectral approach to the problem of face recognition by the fusion of thermal infrared and visual band images. In Yi (2006), by choosing appropriate weights of wavelet transformation coefficients, a novel pixel-level wavelet-based data fusion method is proposed. In Chang et al. (2006), a novel physics-based fusion of multispectral images within the visual spectra is proposed for the purpose of improving face recognition under constant or varying illumination. Spectral images are fused according to the physics properties of the imaging system, including illumination, spectral response of the camera, and spectral reflectance of skin. In Chang et al. (2010), several novel image fusion approaches for spectral face images, including physics-based weighted fusion, illumination adjustment, and rank-based decision-level fusion, are proposed for improving face recognition performance compared to conventional images. A new MSI system is briefly presented which can acquire continuous spectral face images. Singh et al. (2008a, b) present a two-level hierarchical fusion of face images captured under visible and infrared light spectrum to improve the performance of face recognition. At image-level fusion, two face images from different spectrums are fused using DWT-based fusion algorithm. At feature-level fusion, the amplitude and phase features are extracted from the fused image using 2-D Log-Gabor wavelet.

A few of works try to identify the optimal feature band for multispectral face recognition. In Koschan et al. (2011), the fundamentals of MSI and its applications to face recognition are introduced. Then, a complexity-guided distance-based spectral band selection algorithm, which uses a model selection criterion for an automatic selection, is developed to choose the optimal band images under given illumination conditions.

2.2.4 *Multispectral Palmprint*

Multispectral analysis has been used in palm-related authentication (Hao et al. 2007, 2008; Rowe et al. 2007; Likforman-Sulem et al. 2007; Wang et al. 2008a, b). Rowe et al. (2007) proposed a multispectral whole-hand biometrics system. The object of this system was to collect palmprint information with clear fingerprint features, and the imaging resolution was set to 500 dpi. Likforman-Sulem et al. (2007) used multispectral images in a multimodal authentication system. Their system used an optical desktop scanner and a thermal camera which make the system very costly. The imaging resolution is also very high (600 dpi, the FBI fingerprint standard). Wang et al. (2008a, b) proposed a palmprint and palm vein fusion system, which could acquire two kinds of images simultaneously. The system uses one color camera and one near-infrared camera. Hao et al. (2007, 2008) developed a contact-free multispectral palm sensor. Overall, multispectral palmprint scanning is a relatively new topic.

The information presented by multiple biometrics measures can be consolidated at four levels: image level, feature level, matching score level, and decision level (Ross et al. 2006a, b). Wang et al. (2008a, b) fused palmprint and palm vein images by using a novel edge-preserving and contrast-enhancing wavelet fusion method for the use of personal recognition system. Hao et al. (2007) evaluated several well-known image-level fusion schemes for multispectral palm images. Hao et al. (2008) extended their work to a larger database and proposed a new feature-level registration method for image fusion. The results by various image fusion methods were also improved. Although image and feature-level fusion can integrate the information provided by each spectral band, the required registration procedure is often too time-consuming (Wang et al. 2008a, b). As to matching score fusion and decision-level fusion, it has been found (Ross et al. 2006a, b) that the former works better than the later because match scores contain more information about the input pattern and it is easy to access and combine the scores generated by different matchers. For these reasons, information fusion at score level is the most commonly used approach in multimodal biometrics systems and multispectral palmprint systems (Rowe et al. 2007; Likforman-Sulem et al. 2007).

2.2.5 *Multispectral Dorsal Hand*

Since dorsal hand took a role as a biometrics feature in 1990s, the vein underneath the skin has always expressed its good property of permanence and particularity. In traditional research, vein is taken as a web-like structure and most of studies use structure-based feature extraction. However, relevant work shows that even very limited number of minutia missing would cause great performance degradation because the total number of minutiae is usually very small and it is uncomparable to that of other biometrical features (Wang et al. 2008a, b). To pursue higher and more stable results, reserving sufficient original information of vein shape seems to be increasingly important. The original information can be further transferred to various kinds of descriptors by coding or space transformation.

Shape-based feature extraction puts forward higher requirement for original image. Specifically, the vein edge should be clear to avoid the occurrences of broken and blurred vein; the non-vein region should not be extracted as foreground object in case of unwanted interference. Capturing high-quality image has been one of the main guarantees for correct implementation of recognition. In the view of skin optics, light with a different wavelength has a different ability to penetrate skin surface (Aravind and Gladimir 2004). The reason is that various biological tissues in different skin layers vary on their absorptivity and reflectivity. These metrics do not maintain the same values when the wavelength of incident light changes.

Short-wave near-infrared light (700–1100 nm) is widely used in dorsal hand vein capture system, on the ground that deoxyhemoglobin in vein has remarkably higher absorptivity than other tissues in this spectral region. Nevertheless, chromophore, melanin, carotene, and even adipose may bring about negative impact when the light spectrum is closer to red light or long-wave NIR. For example, melanin can severely impede the light flow with shorter wavelength and cause decrease of image contrast. To our knowledge, multispectral dorsal hand study is not well studied. For example, although 850 nm is the most widely used wavelength according to subjective assessment on image quality (Chen et al. 2007), light source optimization has not ever been studied systematically.

2.3 Security Applications

Biometrics applications span a wide range of vertical markets, including security, financial/banking, access control, healthcare, and government applications. Biometrics can be used in both customer- and employee-oriented applications such as ATMs, airports, and time attendance management with the goals of improving the workflows and eliminating fraud.

It is expected that the use of multispectral identification systems to supplement or even replace existing services and methods in some applications with high security requirement, such as border control, citizen ID program, banking, and military.

Border Control
Passengers going aboard or entering a country must present passports and other border-crossing documents to the border guard. It will spend time for the border guard to verify these documents. In order to let border control becoming faster, convenient, and safer, now there are more and more countries start using biometrics passport, such as USA, Canada, Australia, Japan, and Hong Kong. With the development of multispectral biometrics technologies and system, we believe they will play an important role in border control for their high accuracy.

Citizen ID Program
It is a trend for governments to use biometrics technology on the issuance of citizen identity cards. In Hong Kong, a project called Smart Identity Card System (SMARTICS) uses the fingerprint as the authentication identifier. Efficient government services using SMARTICS will provide increased security and faster processing times on different operations such as driver license or border crossing. We think that multispectral biometrics technologies are effective to be used on similar applications.

Banking
The internal operation of banking such as daily authentication process can be replaced by using biometrics technology. Some banks have implemented an authorization mechanism for a different hierarchy of staff by swiping their badge for audit trail purpose. But a supervisor's badge may be stolen, loaned to other

members of staff or even lost. Biometrics system eliminates these kinds of problems by placing an identification device on each supervisor's desk. When a junior member of staff has a request, it is transmitted to the supervisors' computer for biometrics approval and automatically recorded.

Military
Department of Defense of USA distributed more than 11 million Common Access Cards (CAC) as its primary form of identification and enhanced protection to the military network. Although the CAC has proved to be a valuable tool, there are still security gap concerns if cards are lost or stolen and corresponding personal identification numbers are cracked. To fill that void, the Air Force is using biometrics as a way to provide positive identification and authentication (Biometric Technology Working for Military Network 2008). Biometrics is also being used in support of the war on terrorism. Combined with other security measures, biometrics has fast become the preferred solution to military-controlled access and can keep track of who has entered to particular areas because biometrics cannot be shared or borrowed.

2.4 Summary

In this chapter, the MSI technologies have been discussed. Different feature extraction technologies and systems for multispectral biometrics are also discussed. Thus, we have some preliminary understanding of the multispectral biometrics recognition technologies. In the following chapters, the multispectral iris system, multispectral palmprint system, and multispectral dorsal hand system will be presented separately.

References

Akhloufi MA, Abdelhakim B (2010) Locally adaptive texture features for multispectral face recognition. In: IEEE International conference on systems man and cybernetics (SMC), 2010

Aravind K, Gladimir VG (2004) A study on skin optics. Technical report, University of Waterlo, Canada

Bendada A, Moulay AA (2010) Multispectral face recognition in texture space. In: Computer and robot vision (CRV), 2010 Canadian conference

Biometric Technology Working For Military Network (2008) http://americancityandcounty.com/security/military-using-biometrics-0221

Boothapati S, Natalia AS (2013) Encoding and selecting features for boosted multispectral face recognition: matching SWIR versus color. In: SPIE defense, security, and sensing. International society for optics and photonics

Bourlai T, Bojan C (2012) Multi-spectral face recognition: identification of people in difficult environments. In: Intelligence and security informatics (ISI), 2012 IEEE international conference

Boyce CK (2006) Multispectral iris recognition analysis: techniques and evaluation. West Virginia University, pp 101–102

Buddharaju P, Pavlidis I (2007) Multispectral face recognition: fusion of visual imagery with physiological information. Face biometrics for personal identification. Springer, Berlin Heidelberg, pp 91–108

Burge MJ, Monaco MK (2009) Multispectral iris fusion for enhancement, interoperability, and cross wavelength matching. In: Proceedings of SPIE, vol 7334, 73341D

Chang H, Koschan A, Abidi B, Abidi M (2006) Physics-based fusion of multispectral data for improved face recognition. In: 18th international conference pattern recognition, 2006. ICPR 2006

Chang H, Koschan A, Abidi B, Abidi M (2010) Fusing continuous spectral images for face recognition under indoor and outdoor illuminants. Mach Vis Appl 21(2):201–215

Chen L, Zheng H, Li L, Xie P, Liu S (2007) Near-infrared dorsal hand vein image segmentation by local thresholding using grayscale morphology. In: 1st international conference bioinformatics and biomedical engineering

Di W, Zhang L, Zhang D, Pan Q (2010) Studies on hyperspectral face recognition in visible spectrum with feature band selection. IEEE Trans Syst Man Cybern-Part A: Syst Hum 40 (6):1354–1361

Fakourfar H, Belongie S (2009) Fingerprint recognition system performance in the maritime environment. In: Applications of computer vision (WACV), 2009 Workshop

Filter Wheel (2014) http://www.scitec.uk.com/fibreoptics/fw2000.php. Accessed 30 Nov 2014

Hao Y, Sun Z, Tan T (2007) Comparative studies on multispectral palm image fusion for biometrics. In: Asian conference on computer vision, pp 12–21

Hao Y, Sun Z, Tan T, Ren C (2008) Multispectral palm image fusion for accurate contact-free palmprint recognition, In: International conference on image processing, pp 281–284

Hyperspectral imaging (2014) http://en.wikipedia.org/wiki/Hyperspectral_imaging. Accessed 30 Nov 2014

Khalil MS, Muhammad D, AL-Nuzaili Q (2009) Fingerprint verification using the texture of fingerprint image. In: Second international conference, machine vision, 2009. ICMV'09

Koschan A, Yao Y, Chang H, Abidi M (2011) Multispectral face imaging and analysis. Handbook of face recognition. Springer, London, pp 401–428

Likforman-Sulem L, Salicetti S, Dittmann J, Ortega-Garcia J, Pavesic N, Gluhchev G, Ribaric S, Sankur B (2007) Final report on the jointly executed research carried out on signature, hand and other modalities. http://www.cilab.upf.edu/biosecure1/public_docs_deli/BioSecure_Deliverable_D07-4-4_b2.pdf.pdf

Liquid Crystal Tunable Filter (2014) http://en.wikipedia.org/wiki/Liquid_crystal_tunable_filter. Accessed 30 Nov 2014

Liu Z, Yan J, Zhang D, Li Q (2007) Automated tongue segmentation in hyperspectral images for medicine. Appl Opt 46:8328–8334

Nicolo F, Natalia AS (2011) A method for robust multispectral face recognition. Image analysis and recognition. Springer, Berlin Heidelberg, pp 180–190

Park J, Kang M (2007) Multispectral iris authentication system against counterfeit attack using gradient-based image fusion. Opt Eng 46:117003

Ross A, Pasula R, Hornak L (2006a) Exploring multispectral Iris recognition beyond 900 nm. In: Proceedings of the 2006 conference on computer vision and pattern recognition workshop: 51

Ross AA, Nadakumar K, Jain AK (2006b) Handbook of multibiometrics, Springer, Berlin

Rowe RK, Nixon K, Corcoran S (2005) Multispectral fingerprint biometrics. In: Proceedings from the sixth annual IEEE SMC, Information assurance workshop, 2005. IAW'05

Rowe RK, Uludag U, Demirkus M, Parthasaradhi S, Jain AK (2007) A multispectral whole-hand biometric authentication system. In: Biometrics Symposium, pp 1–6

Rowe RK, Nixon KA, Butler PW (2008) Multispectral fingerprint image acquisition. Advances in biometrics. Springer, Berlin, pp 3–23

Singh R, Vatsa M, Noore A (2008a) Multiclass mv-granular soft support vector machine: a case study in dynamic classifier selection for multispectral face recognition. In: 19th international conference on pattern recognition, 2008. ICPR 2008

Singh R, Vatsa M, Noore A (2008b) Hierarchical fusion of multi-spectral face images for improved recognition performance. Inf Fusion 9(2):200–210

Wang J, Yau W, Suwandy A, Sung E (2008a) Person recognition by fusing palmprint and palm vein images based on "Laplacianpalm" representation. Pattern Recogn 41(5):1514–1527

Wang L, Leedham G, Cho DS-Y (2008b) Minutiae feature analysis for infrared hand vein pattern biometrics. Pattern Recogn 41:920–929

Wilkerson CL, Syed NA, Fisher MR, Robinson NL, Wallow IHL, Albert DM (1996) Melanocytes and iris color: light-microscopic findings. Arch Ophthalmol 114:437–442

Yi M (2006) Multispectral imaging for illumination invariant face recognition

Zheng Y (2011) Orientation-based face recognition using multispectral imagery and score fusion. Opt Eng 50(11): 117202

Zheng Y, Elmagbraby A (2011) A brief survey on multispectral face recognition and multimodal score fusion. In: Signal processing and information technology (ISSPIT), 2011 IEEE international symposium

Zheng Y, Zhang C, Zhou Z (2012) A wavelet-based method for multispectral face recognition. In: SPIE defense, security, and sensing, international society for optics and photonics

Part II
Multispectral Iris Recognition

Chapter 3
Multispectral Iris Acquisition System

Abstract Multispectral iris recognition is one of the most reliable biometrics in terms of recognition performance. This paper describes the design and implementation of a high-speed multispectral iris capture device, which consists of the following four parts: (1) capture unit; (2) illumination unit; (3) interaction unit; and (4) control unit. A multispectral iris image database is created by the proposed capture device, and then, we use the iris image-level fusion to further investigate the effectiveness of the proposed capture device by the 1-D Log-Gabor wavelet filter approach.

Keywords Multispectral iris · Acquisition system · Fusion · Recognition

3.1 System Requirements

Biometrics has become more important, with an increasing demand on security. Iris recognition is one of the most reliable and accurate biometric technologies in terms of identification and verification performance. It mainly uses iris patterns to recognize and distinguish individuals since the pattern variability among different persons is enormous. In addition, as an internal organ of the eye, the iris is well protected from the environment and is stable over time. The amount of the iris texture information will greatly affect the performance of the recognition algorithm.

A critical step in an iris recognition system is designing an iris capture device that can capture iris images in a short time. Some research groups (Wildes 1997; Park et al. 2005; Tan et al. 1999; CASIA Iris Image Database 2005; Shi et al. 2003), such as OKI, LG, Panasonic, and Cross-match, have explored the requirements on the iris image acquisition system, and some implementations have already been put into commercial practice (Biom Technol 2005; Mobile Dual Iris Capture Device 2014; Oki 2002; Iris Recognition 2014; Iris Recognition Camera System 2009). The Institute of Image Processing and Pattern Recognition at Shanghai Jiao Tong

D. Zhang et al., *Multispectral Biometrics*,
DOI 10.1007/978-3-319-22485-5_3

University also developed a contactless auto-feedback iris capture system (He et al. 2008).

The previous different techniques have their limitations. All these capture devices operate predominately in single band of the near-infrared (NIR) range of the electromagnetic spectrum, using wavelengths that peak around 850 nm, with a narrow band pass. The wavelength of 850 nm has some strength, such as alleviating physical discomfort from illumination, reducing specular reflections, and increasing the amount of texture captured for some iris colors. Commercial iris recognition systems operate in single band of the NIR for another primary reason: simplifying system design and saving reducing production costs.

But the textural information of iris has the complex components, mainly including two kinds of texture: from structures and from pigments. The wavelength of 850 nm can penetrate the pigments, presenting the texture which cannot be easily observed in the visible spectrum. But the effect of pigments, the major color-inducing compound, is negligible at 850 nm. So the iris images captured by the previous devices have insufficient textural information, lacking of the ones generated by the pigments.

The above study has inspired researchers into considering that the iris textures generated outside 850 nm may have more information over those that are only generated in the NIR spectrum, because pigments can be present in the shorter wavelengths and become another major source of iris texture.

Previous research has shown that matching performance is not invariant to iris color and can be improved by imaging outside the NIR spectrum and that the physiological properties of the iris (e.g., the amount and distribution of pigments) impact the transmission, absorbance, and reflectance of different portions of the electromagnetic spectrum and the ability to image well-defined iris textures. Performing multispectral fusion at the score level is proven feasible, and the multispectral information is used to determine the authenticity of the imaged iris (Wilkerson et al. 1996).

Some research groups (Ross et al. 2006; Vilaseca et al. 2008; Burge et al. 2009; Ngo et al. 2009) explored the multispectral iris image capture devices, see Table 3.1. Most of the previous multispectral devices switch the light source or filter manually, adjust the lens focal length manually, and use chin rest to require an uncomfortable fixed head position. To some extent, most of the multispectral devices demand full cooperation from the subject who needs to be trained in advance, which will eventually increase the time of image acquisition and influence the acceptability of users. Due to the time-wasting capture process, subjects may be tired and easily fatigued, blinking eyes, rotating eyes, and dilating and constricting pupil subconsciously, which are interference factors to iris image quality and influence the accuracy of the recognition. So, due to three disadvantages, inefficiency, full user cooperation, and low quality, the previous multispectral iris capture devices are designed for the experimental data collection, far from the requirements of real usage scenarios.

We presented a high-speed multispectral iris capture system that will enable collection of data in a short time, to explore the feasibility of multispectral capture

Table 3.1 The comparison of previous multispectral devices

	Designer	Wavelength switching	Focusing mode	Capturing mode	Capturing speed	Cooperation	Interaction	Recognition
1	West Virginia University	Manually	Manually	Manually	Slow	High	No	No
2	Technical University of Catalonia, The University of Granada	Manually	Manually	Manually	Slow	High	No	No
3	Security and Intelligence, USA	Manually	Manually	Uncertain	Slow	High	No	No
4	United States Naval Academy, USA	Automatic	Manually	Manually	Medium	High	No	No

system with efficiency and user-friendliness. Using this system, a complete capture cycle (including three wavelengths) can be completed within 2 or 3 s, much faster than above devices. In addition, this system is not only an isolated acquisition device, but also connects to the server running recognition algorithm, so it can complete a full process of multispectral online iris identification, which is the first attempt of practical application of the multispectral iris recognition.

The capture system consists of the following four parts: (1) capture unit; (2) illumination unit; (3) interaction unit; and (4) control unit. It uses a SONY CCD (charge-coupled device) camera and an automatic focusing lens as the capture unit, and the working distance is about 300 mm. Two groups of matrix-arrayed LEDs (light-emitting diodes) across three different wavelengths (including visible light and near-infrared light) are used as the main multispectral illumination unit, and three LEDs in triangular arrangement are used as the auxiliary illumination unit, specially designed for pupil location. We design an interaction unit including the infrared distance measuring sensor and the speaker to realize the exact focusing range of the lens via the real-time feedback of the voice prompts. The novel design on the control unit synchronizes the previous three units and makes it capturing with high speed, easy to use, and nonintrusive for users. The system is designed to have good performance at a reasonable price so that it becomes suitable for civilian personal identification applications.

3.2 Parameter Selection

One of the major challenges of multispectral iris capture system is capturing an iris image of the iris while switching wavelength band of illumination. The realization of the capture device is quite complicated, for it integrates light, machine, and electronics into one platform and involves multiple processes of design and manufacture.

A multispectral iris image capture system is proposed. The design of the capture device includes following four subcomponents: (1) capture unit; (2) illumination unit; (3) interaction unit; and (4) control unit. The capture, illumination, and interaction units constitute the main body of the capture device, which is installed on the 3-way pan-tilt head of a tripod to allow subjects to manually adjust the pitching angle for fitting their height while capturing. The control unit is operating within the capture system, in charge of the lens focusing, synchronization of other three units, and data exchange with the iris recognition server.

We considered the configurations for our system as flowing: use one single camera with multiple narrow band illuminators. The illuminators are controlled by ARM (Advanced RISC Machines) main-board via a single chip sub-board and can switch automatically and synchronize with the lens focus and CCD shutter, as shown in Fig. 3.1. This approach enables high-speed collection of image. Using this system, in theory, the acquisition speed is limited only by two factors: the frame rate of camera and the switching speed of multispectral illuminators. In fact, in

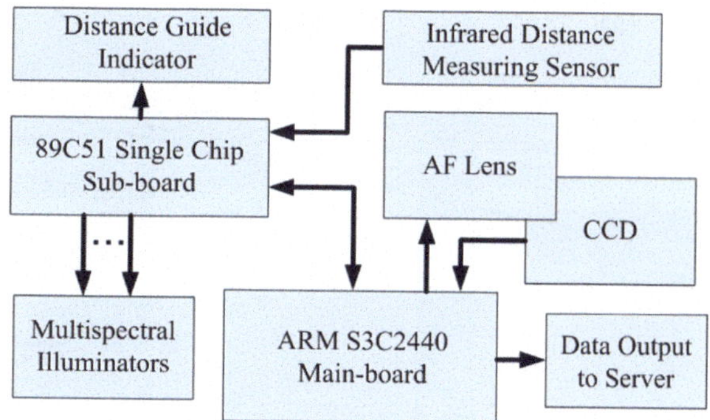

Fig. 3.1 Block diagram of a multispectral iris acquisition system

order both to maintain high-speed acquisition and to ensure image quality, some measures are taken to eliminate the interference factors, which make the actual collection rate lower than the theoretical.

The optical path is as follows: First, the subject is watching the reflection filter, and the multispectral light from the matrix-arrayed illuminators is delivered to the eye, and then, the reflected light from the subject's eye is collected through the filter, through the close-up lens, and imaged by the ICX205 camera using AF lens. In addition, the light beams from IR distance sensor and red-eye LEDs are also concentrated on the surface of subject's iris. The optical path of the proposed capture device is shown in Fig. 3.2a, b.

3.2.1 Capture Unit

The capture unit that we propose is composed of 5 parts: Sony ICX205 HAD (hole accumulation diode) CCD sensor, AF lens, close-up lens, reflection filter, and lens hood, as shown in Fig. 3.3. The camera with the Sony ICX205 HAD CCD sensor is using USB interfaces and has exceptional features, including high resolution (working at 640 × 480), high sensitivity (maximum sensitivity = 0.01Lux at F1.2), and low dark current (S/N ratio > 60 dB), which are all important to multispectral imaging. The CCD spectral response from 400 to 1000 nm wavelengths, including visible light and infrared light, is not absolutely uniform, as shown in Fig. 3.4, but we verified that the CCD response does not introduce significant errors into the experimental values after the optimization of the multispectral system.

In iris capture system, the acquisition of iris images almost always begins in poor focus. It is therefore desirable to compute focus scores for image frames very rapidly, to control a moving lens element for auto-focusing. AF lens is DC driven and controlled by the ARM main-board, which running the 2-D fast focus assessment algorithm to estimate the quality of focus of a multispectral image and to indicate the direction of lens focus's movement. The AF lens has a relatively large minimum focusing distance, generally not less than 500 mm, which will limit the

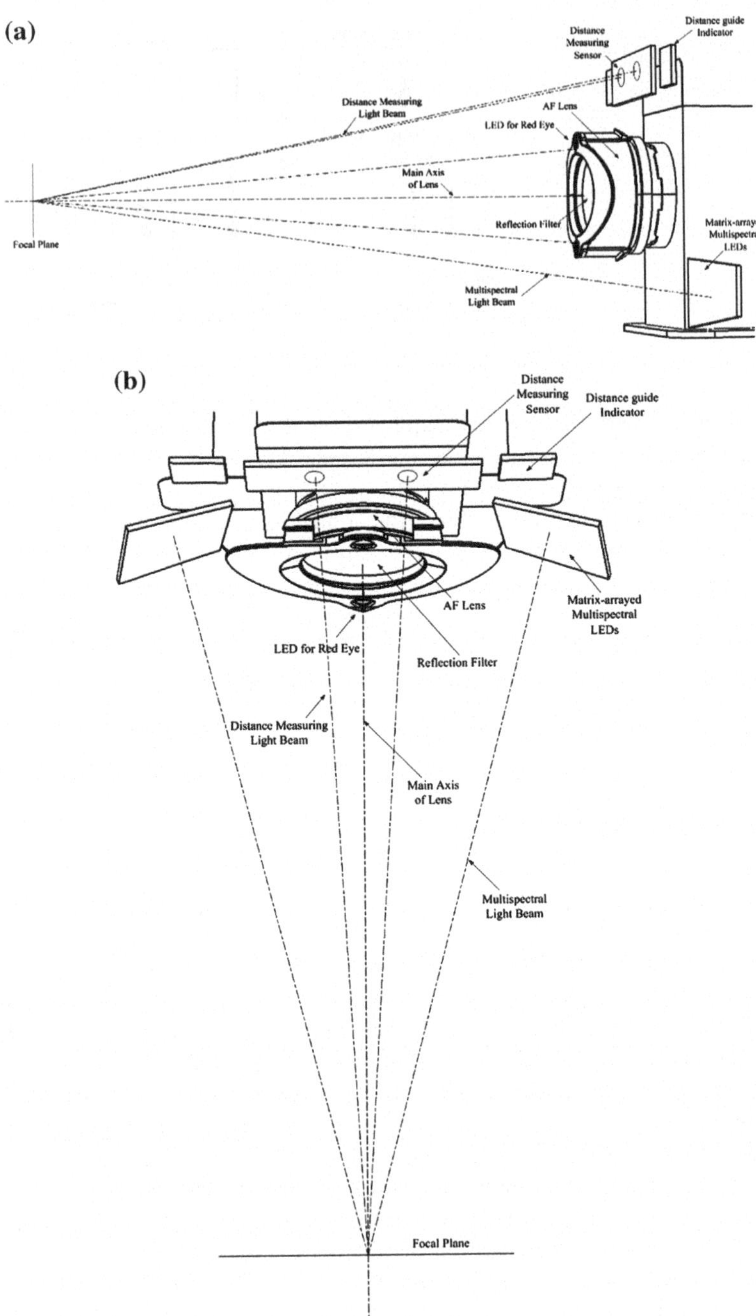

Fig. 3.2 The optical path. **a** Side view, **b** Front view

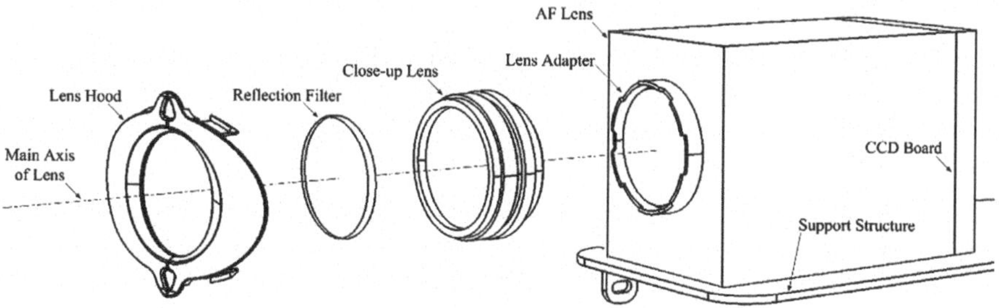

Fig. 3.3 The composition of capture unit

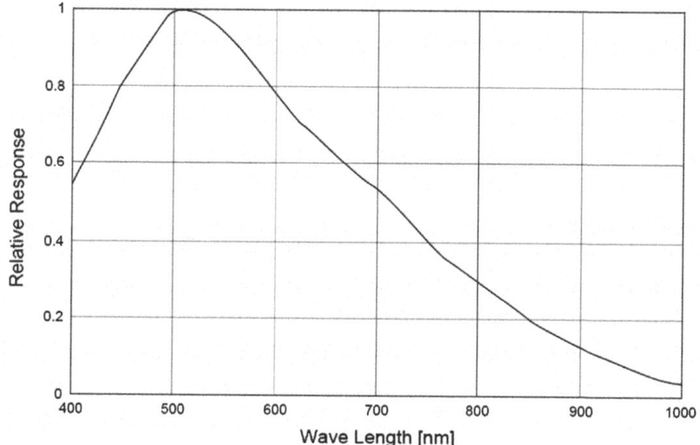

Fig. 3.4 Relative sensor response for the Sony ICX205

iris capture. The close-up lens allows the AF lens to be focused at a much closer distance—resulting in a higher magnification of the subject. The quality of the close-up lens is very good: the image to be reasonably sharp in the center, and a little less sharp toward the edges, which will not have a serious impact on image quality. With closing-up lens, the focus range of the AF lens is set from 200 to 300 mm, so we can capture iris image that is in acceptably sharp focus in this distance range, without the strict requirements on the subject's location and cooperation.

The reflection filter is the customized coated filter corresponding to the wavelengths of illuminators. The reflection filter can transmit the light of specified three wavelengths and reflect the light of other wavelengths, as shown in Fig. 3.5. These three wavelengths with high transmittance are 700, 780, and 850 nm, under which the iris is imaged by CCD, and the reflection of all other wavelengths looks like a mirror. When subject is watching the mirror image of eye in the filter, his eye is just on the main axis of the lens, so the iris will be located in the center of the image.

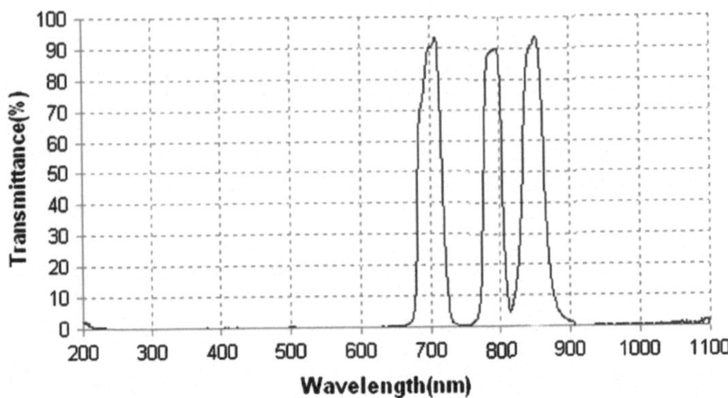

Fig. 3.5 The transmittance of the reflection filter across all wavelengths

Lens hood is used on the end of reflection filter to block the surrounding ambient light or other light source in order to prevent glare and lens flare, and to ensure that only the reflected light of three specific wavelengths (700, 780, and 850 nm) can enter the lens vertically.

The output of capture unit is the sequence of multispectral iris images received by CCD. The images are transmitted using two routes: One is from the CCD to the ARM main-board, which can run the 2-D fast focus assessment algorithm to estimate the focus quality of image for lens auto-focusing, and the other is from the ARM main-board to server (the image processing host computer) for iris recognition.

3.2.2 *Illumination Unit*

The illumination system has two parts: the main illuminators and the auxiliary ones. The main illuminators are composed of two groups, which are located at the bottom side of the lens, and each group is including of nine matrix-arrayed (3 × 3) LEDs corresponding to three wavelengths: 700, 780, and 850 nm. The wavelengths of main illuminators can be switched automatically, allowing illumination of the captured iris with a 70° angle. The auxiliary illuminators are composed of two groups, which are located above and below the AF lens, and each group is including three LEDs corresponding to three wavelengths: 700, 780, and 850 nm, allowing illumination of the captured iris with an almost 90° angle, as shown in Fig. 3.6. The combination of above two kinds of illuminators from two directions is especially designed for accurate pupil location in multispectral images, which will influence the performance of iris recognition greatly.

When the iris is illuminated by a light source, the light enters the pupil and is reflected off the retina and comes back to the light source again through the pupil

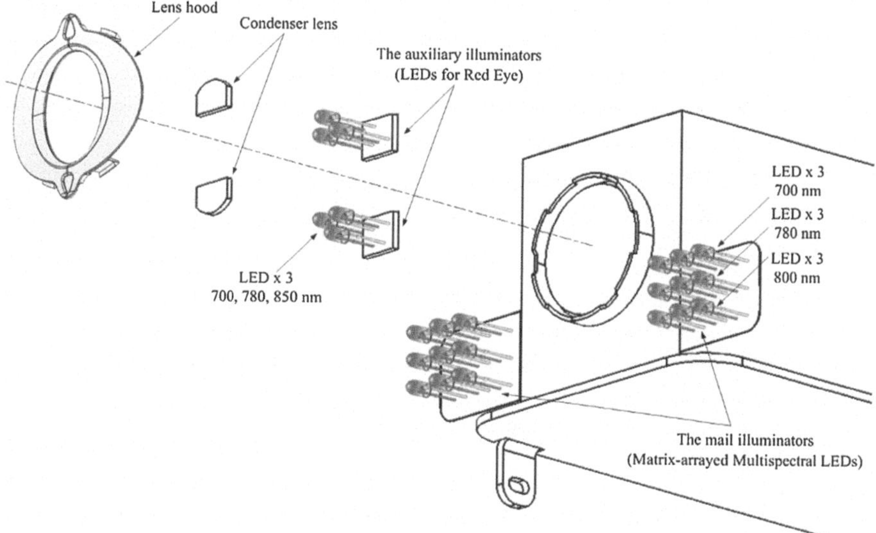

Fig. 3.6 The composition of illumination unit

because of the property of the optical system of the eyeball (Ebisawa 1998). Therefore, if the light source is set coaxial with the lens, the pupil in image appears as a half-lighted disk against a darker iris background, called the bright pupil or "red eye", as shown in Fig. 3.7a, b. In contrast, if the eye is illuminated by a light source uncoaxial with the lens, the pupil in image appears as a darker area against the iris background, called the dark pupil, as shown in Fig. 3.7c, d. As shown in

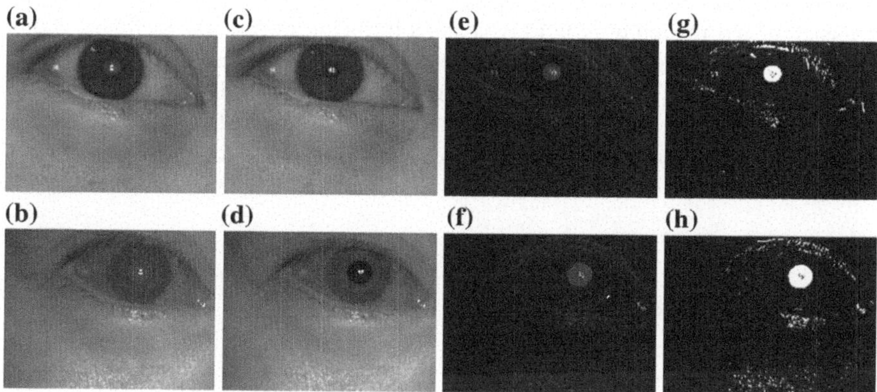

Fig. 3.7 The *bright* and *dark* pupil images under different wavelengths: **a** and **c** Images captured under 700 nm from the same iris, with *bright* and *dark* pupil, respectively, **b** and **d** Images captured under 850 nm from the same iris, with *bright* and *dark* pupil, respectively, **e** Difference images obtained by subtracting (**c**) from (**a**), **f** Difference images obtained by subtracting (**d**) from (**b**), **g** Binary image from (**e**), **h** Binary image from (**f**)

Fig. 3.7a–d, in image of some spectral wavelengths, whether under bright or dark pupil condition, the gray level of pupil seems too similar with the ones of iris to be segmented from each other.

To solve this problem, in the proposed design, the main illuminators set uncoaxial with the lens are switched on during the odd frames, and the auxiliary ones set coaxial with the lens are switched on during the even frames (the sequence of iris images consists of one odd frame and one even frame). The difference images are obtained by subtracting the even frame images from the consecutive odd frame images. As a result, the iris area almost vanishes as shown in Fig. 3.7e, f, and the pupil area will be more obvious after dynamic thresholding, as shown in Fig. 3.7g, h. Here, it is necessary to make the brightness levels of iris area in the odd and even frame images equal by controlling the illuminators' current. After binarizing the obtained difference images with a dynamic or preset threshold, an approximate region of the pupil area can be detected. Using the approximate region as reference, we can locate and segment the pupil in bright or dark ones. This design of illumination helps to improve the accuracy of pupil detection across all spectral wavelengths.

To avoid glittering, we optimized two kinds of illuminators: The arrangement and the angle of each LED are specially designed. Therefore, the glittering is controlled in the pupil area in most cases, and this glittering does not affect the localization and recognition. The selection of above three wavelengths is mainly based on two reasons. The first reason is that we found three clusters that are enough to present all wavelengths, including the visible and infrared spectrum. The selection of three wavelengths 700, 780, and 850 nm is the optimized result, in order to ensure adequate coverage of the full spectrum and diversity of the iris texture. The second reason is the practical considerations. The proposed iris capture device in this work will be developed into the practical multispectral iris recognition system, not just the experimental device, so all the selected wavelengths should be easily accessible in system. Selection of more wavelengths maybe means a little more accurate performance of iris recognition, but also results in much longer time of image acquisition and much lower acceptability of users. In summary, the selection of above three wavelengths is the trade-off between the experimental data integrity and the feasibility of the practical system. In addition, the luminance levels for the experimental setup described here meet the requirements found in the ANSI/IESNA RP-27.1-05 for exposure limits under the condition of weak aversion stimulus.

3.2.3 Interaction Unit

The interaction unit is included in our iris capture device system for the purpose of easily capturing iris images. It is very convenient for subjects to adjust their poses according to the feedback information. The interaction unit that we propose is composed of 4 parts: reflection filter, infrared distance measuring sensor, distance guide indicator, and the speaker, as shown in Fig. 3.8.

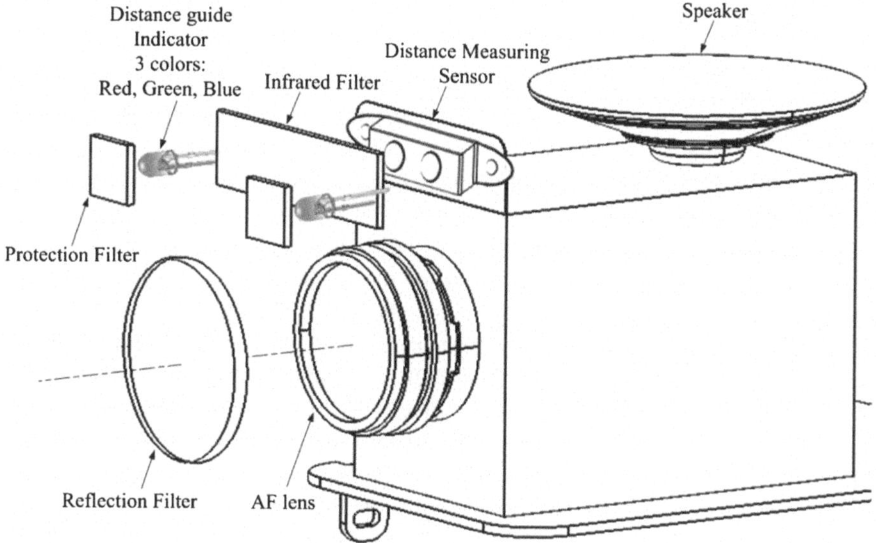

Fig. 3.8 The composition of interaction unit

As previously mentioned, under the visible wavelengths shorter than 700 nm, the reflection filter has the mirror effect for the subject to stare at the mirror image of eye in the center. The reflection filter can ensure that the subject's eye is just on the main axis of the lens; in other words, the iris is located in the center of the image. Using reflection filter, eye position is less likely to be out of site, making eye alignment easier.

The infrared distance measuring sensor is GP2D12 of Sharp, with integrated signal processing and analog voltage output. The measuring distance range is from 100 to 800 mm. When the infrared distance measuring sensor gets the distance between the subject and the capture unit, both the distance indicator and the speaker provide subject guidance so the correct capture position is achieved. According to the focus range of AF lens (from 200 to 300 mm), if the distance is out of the above range, the distance indicator will blink with red or blue LEDs and the voice instructions will be "Please move back" or "Please move closer".

In effect, subjects can make quick adjustment on their poses according to the guidance information, including light and voice. When the capture is completed, the distance indicator will automatically blink with green LEDs to tell the subject that it is successful in capturing a good iris image.

The three units' main light axes (including capture, illumination, and distance measuring) intersect at the point that is 250 mm from the AF lens, just at the middle of focus range (200–300 mm).

3.2.4 Control Unit

The control unit is composed of an ARM S3C6410 main-board and an 89C51 single chip sub-board. The former is running Windows CE 5.0, can make fast focus assessment of image, drive AF lens, and synchronize the previous three units: capture, illumination, and interaction. The latter controls distance sensor, guide indicator, and illuminators.

In iris capture system, the acquisition of iris images is almost always interfered by pupil dilation. It is difficult to perform the accurate image registration and image-level fusion on the normalized iris images with different pupil diameters. It is therefore desirable to capture a sequence of iris images with the same or similar pupil diameter, to introduce as little interference to multispectral fusion as possible. To control pupil dilation, we can use AF mode of lens and max frame rate of CCD and capture continuously a sequence of iris images under different wavelength illumination switched automatically with high speed.

A complete cycle of iris dilation will take some time, about 1.5–2 s, so the speed of capture faster, the degree of iris dilation smaller. There are three factors influencing capture speed: the switching speed of wavelengths, the frame rate of CCD, and the focusing speed of lens. The switching time of wavelengths is small enough to be negligible, and the frame rate is limited by the bandwidth of CCD, almost no room for improvement. So the focusing speed is the most important factor to iris dilation.

The refractive index of the same lens changed with the spectral wavelength, so the lens focus should be adjusted each time that one wavelength illumination was switched on. Among the 3 wavelengths, 700 nm is the shortest, with the largest refractive index and minimum object distance; 850 nm is the longest, with the smallest refractive index and maximum object distance; and 780 nm is between them. The lens focus in traditional capture device is always manually adjusted by operator, and it is impossible to meet the requirement of high-speed capture, so the AF lens is necessary in this design.

The focusing speed is limited by two factors: the efficiency of focusing algorithm and the mechanical movement of lens. The latter almost cannot be changed, while the former has much room for improvement. We use the improved convolution matrix method as the focusing algorithm. Daugman's original method (Daugman 2004) sometimes produces incorrect quality scores when working with iris images. Since it computes the first and second derivatives of neighboring pixels, the presence of eyelashes often causes great inconsistencies in quality scores. To overcome this problem, we use a new (5×5) convolution kernel (Kang et al. 2007) as shown in Fig. 3.9a, and the power spectrum is shown in Fig. 3.9b.

It is time-consuming to run the focusing algorithm in the entire image, so we select a set of continuous subregions with the 5×5 size in the middle of the image. Although there are always many interference factors in the image, such as eyelashes, eyelids, and skin, but the cross-area near the center of the image is more likely to be the iris, the most important object to be focused clearly.

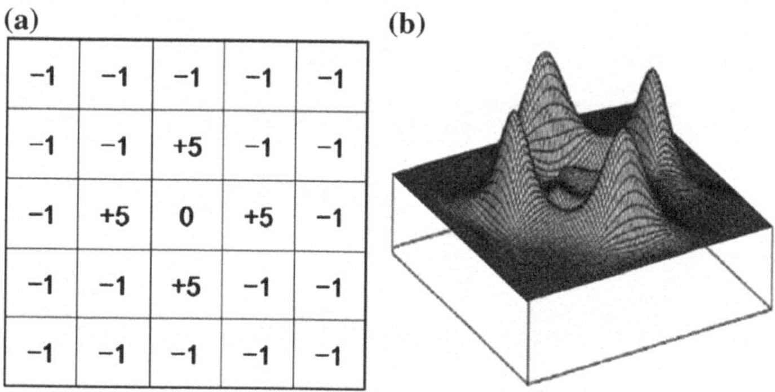

−1	−1	−1	−1	−1
−1	−1	+5	−1	−1
−1	+5	0	+5	−1
−1	−1	+5	−1	−1
−1	−1	−1	−1	−1

Fig. 3.9 **a** The proposed convolution kernel, **b** the corresponding power spectrum

Suppose we obtain N subregions from the iris image, where n is an integer. For each of the n subregions, we generate a convolution value based on our improved convolution kernel, as described previously. We denote by X_s, the convolution value of the Sth subregion, and by δ_N, the standard deviation σ of N subregions. The Sth subregion has a mask M_S quantized based on the following equation:

$$M_s = \begin{cases} 1 & \text{if} \frac{1}{N}\sum_{S=1}^{N} X_S - 3\delta_N < X_s < \frac{1}{N}\sum_{S=1}^{N} X_S + 3\delta_N \\ 0 & \text{else} \end{cases} \tag{3.1}$$

$\frac{1}{N}\sum_{S=1}^{N} X_s \pm 3\delta_N$ is the preset threshold to remove further noise. So, the focus score for image based on the improved convolution matrix, D_0, is defined as following:

$$D_0 = \frac{1}{\sum_{S=1}^{N} M_S} \sum_{S=1}^{N} (X_S \cdot M_S) \tag{3.2}$$

ARM main-board is running the focusing algorithm to estimate the focus quality of a multispectral image and to indicate the direction of lens focus's movement for higher focus score, until the focus score reaches a preset threshold, and then, the lens is correctly focused. The computational time of focusing algorithm on each image is less than 15 ms using S3C6410 at 533 MHz. Taking into account the mechanical movement of lens, the action cycle of one focusing will take not more than 200 ms.

The control unit manages the working process of one multispectral data collection cycle and synchronizes the other three units: capture, illumination, and interaction, as shown in Fig. 3.10. Once subject moves closer to the capture device, the distance measuring sensor will output the distance voltage signal. The work cycle of distance measuring sensors is 40 ms. When the single chip sub-board

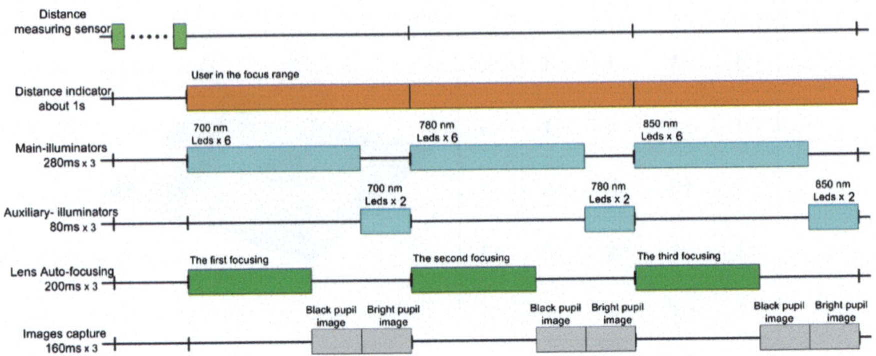

Fig. 3.10 The control signal timing diagram of one multispectral data collection cycle

receives voltage signal and serializes distance value, a drive signal will be sent to the distance guide indicator, instructing subject to move closer or back. If subject moves into the focus range of AF lens, the illumination unit will be turned on in order of the wavelengths: 700, 780, and 850 nm. At every time of one wavelength illumination turns on, the lens begins auto-focusing and CCD begins acquisition. In the process of capture, the main and auxiliary illuminators corresponding to the same wavelengths should be switched on successively to capture the images with bright pupil and dark pupil. Taking into account that it will take some time for subject to move into the focus rage, a complete capture cycle generally can be completed within 2 or 3 s, and then, the images will be transferred to server via USB 2.0 interface.

3.3 System Performance Evaluation

In this section, a series of experiments is performed to evaluate the performance of the proposed multispectral iris capture device. First, a multispectral iris image database is created by the proposed capture device. Then, we use the iris image-level fusion to further investigate the effectiveness of the proposed capture device by the 1-D Log-Gabor wavelet filter approach proposed by Masek (2003).

3.3.1 Proposed Iris Image Capture Device

According to the proposed design, we developed a multispectral iris image capture device, which is shown in Fig. 3.11a, b. The dimension of this device is 130 mm (width) × 130 mm (height) × 180 mm (thickness). The device's working distance is about 200–300 mm. Through a USB 2.0 cable, this device can connect to the server

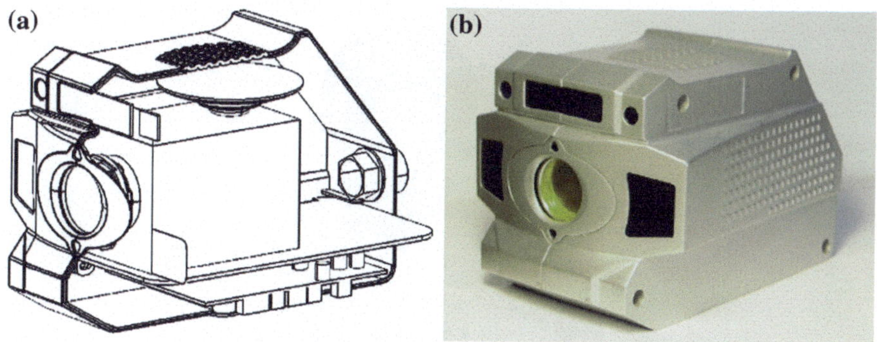

Fig. 3.11 The multispectral iris capture device. **a** The internal structure. **b** The external form

computer running recognition algorithm and complete a full process of multispectral online iris identification.

The light intensity across three wavelengths is different due to the specifications of the illuminators, LEDs. The reflectivity of the iris also varies across three wavelengths. So the brightness of iris area in images collected by the proposed capture device is dependent on both the reflectivity of iris and the light intensity of illuminators. Before the acquisition, CCD is switched to auto-exposure mode and shutter time is preset as 1/30 s, in order to minimize the difference in the brightness of images and to adapt to the max frame rate 30fps of CCD.

For each subject, the capture device will automatically start a multispectral data collection cycle once it ensures that the subject has move into the focus range. In the acquisition, the subjects do not need to do other things except to watch the reflective filter. Each wavelength illuminator is switched on; two images (including one "bright pupil" image and one "dark pupil" image) are captured and transferred to the server, then the current wavelength illuminator is turned off, and the next wavelength illuminator repeats the above process. The scene of the multispectral data collection is shown in Fig. 3.12.

The multispectral iris images that are captured by this device can be seen in Fig. 3.13. As shown in Fig. 3.13a–f, we can know that the captured iris image quality is very good and the pupil radius has a relatively consistent and small size. The glittering is controlled in the pupil area in most cases, and this glittering does not affect the localization and recognition.

3.3.2 *Iris Database*

A dataset that contained samples from 80 irides was used to conduct the following study. Our iris image database, which used the proposed multispectral iris capture device, was created with 40 subjects. In this dataset, 25 are male. The age distribution is as follows: younger than 30 years old comprise 80 % and between 30 and 40 years old comprise about 15 %, see Table 3.2.

Fig. 3.12 The scene of the multispectral data collection

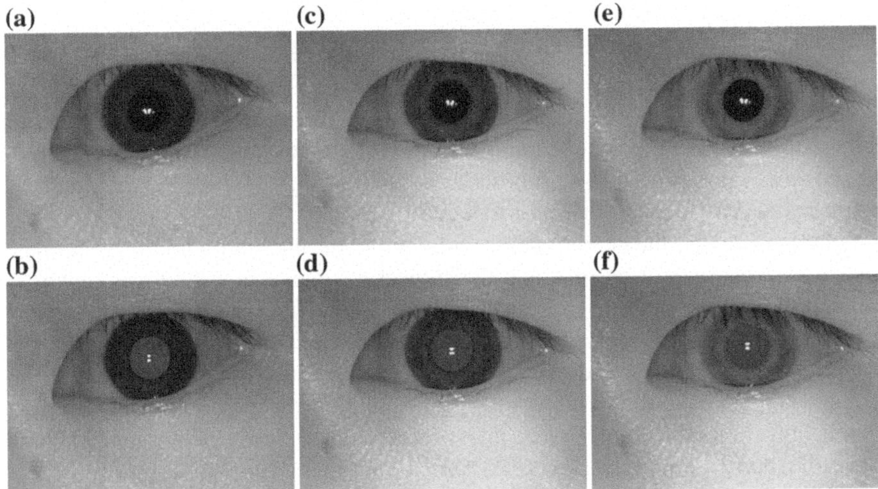

Fig. 3.13 The multispectral iris images: **a** and **b** captured under 700 nm, **c** and **d** captured under 780 nm, **e** and **f** captured under 850 nm, from one multispectral data collection cycle for the same iris

Table 3.2 The composition of the multispectral iris image dataset

Total number of subjects	40
Number of males	25
Number of females	15
Age 21–30	32
Age 31–40	6
Age 41–50	2

The image resolution is 640×480, and the distance between the device and the subject is about 250 mm. Ten pairs of images (twenty images, including ten "bright pupil" images and ten "dark pupil" images), which were taken from each of the left and right eye of a subject, are selected under each of three spectral bands corresponding to 700, 780, and 850 nm, respectively. So, each subject has experienced ten multispectral data collection cycles. In total, we collected 4800 iris images for this database, which are used as a unique session in the experiment.

3.3.3 Image Fusion and Recognition

Using the traditional recognition method, iris images are only matched within single wavelength, and iris codes extracted from different wavelength images cannot be cross-matched to each other. In order to achieve higher accuracy than single wavelength recognition, we presented a new image-level fusion approach for the multispectral iris images.

As mentioned previously, we can locate pupil in iris image based on the image difference method. After segmentation, we can use the homogenous rubber sheet model devised by Daugman (1993) to normalize the iris image, by remapping each pixel within the iris region to a pair of polar coordinates (r, θ) where r is on the interval [0,1] and θ is angle $[0, 2\pi]$. So, we can obtain six normalized patterns from one multispectral data collection cycle.

1-D Log-Gabor wavelet recognition method proposed by Masek (2003) is used for encoding in our experiments. 1-D Log-Gabor band-pass filter can be efficient in angular feature extraction, which is the most distinctive and stable texture information, and ignores the radial feature extraction, which is easily interfered by dilated pupil. 1-D Log-Gabor wavelet method is the most popular comparison method used in the literature due to the accessibility of their source code. We use the performance of 1-D Log-Gabor wavelet method as the benchmark, to access the improvement of recognition performance after multispectral iris image fusion.

1-D Log-Gabor wavelet feature encoding method proposed by Masek is implemented, by convolving the normalized pattern with 1-D Log-Gabor wavelets. The rows of the 2-D normalized pattern (the angular sampling lines) are taken as the 1-D signal, and each row corresponds to a circular ring on the iris region. The angular direction is taken rather than the radial one, which corresponds to columns of the normalized pattern, since the iris maximum independence occurs in the angular feature extraction, which is the most distinctive and stable texture information, and ignores the radial feature extraction, which is easily interfered by dilated pupil. We revisit the default parameter values used by Masek to get higher recognition performance (Peters 2009). All parameters revised in this work are as follows: angular resolution and radial resolution of normalized image, center wavelength and filter bandwidth of 1-D Log-Gabor filter, and fragile bit percentage, see Table 3.3. After we apply the 1-D Log-Gabor wavelet filter to the normalized iris image, we quantize the result to create the iris code by determining the quadrant

Table 3.3 Listing of parameters revised for multispectral iris images with the initial value from Masek

Parameter	Initial value	Revised value
Angular resolution (θ)	240 pixels	360 pixels
Radial resolution (*r*)	20 bands	60 bands
Center wavelength (λ)	18 pixels	16 pixels
Filter bandwidth (σ/*f*)	0.5	0.4
Row averaging	NA	3 rows
Fragile bit percentage	NA	Yes

of the response in the complex plane. This gives us an iris code that has twice as many bits as the normalized iris image had pixels.

Given some multispectral images of the same iris under different wavelength illumination, there will be differences in iris code due to variations in iris texture generated by structures or pigments. Some obvious textures can result in "stable bits" within iris code, but other unobvious textures can result in "fragile bits" within the iris code, such as either a zero or a one with some degree of randomness (Hollingsworth et al. 2009). So, more "fragile bits" mean less stable iris texture features or lower recognition confidence level. By identifying the fragile bits and calculating the percentage of the fragile bits in iris code corresponding to each row of normalized pattern, we can choose the rows of normalized pattern from different wavelengths and fuse them into one image.

We identify fragile bits closest to the axes for the real and imaginary response with the preset threshold, which is calculated using the absolute values of the filter response. Suppose we have B bands of spectral wavelengths, and in one multispectral data collection cycle, S iris images from the same eye of the same subject are captured under each band of spectral wavelengths, so the total number of iris images is S × B. The multispectral image fusion is defined as the following:

$$R_i = \{R_{i,j} | \arg\min f_{\text{percent}}(i,j)\}, \quad i = 1, 2, \ldots, 60, j = 1, 2, \ldots, S \times B \tag{3.3}$$

R_i is the No. i row in normalized pattern of multispectral image fusion, $R_{i,j}$ is the No. i row in normalized pattern of No. j iris image, and the $f_{\text{percent}}(i,j)$ is the percentage of the fragile bits in iris code corresponding to No. i row in normalized pattern of No. j iris image. Using this method, we can select the row which has the minimum number of fragile bits across different wavelengths, and fuse them into one normalized iris pattern.

In our work, six iris images captured under three wavelengths are combined into one normalized pattern. Six images were taken at a very short interval of time (all time-consuming no more than 3 s), generally without obvious iris movement and dilation, so the fusion of these iris normalizes pattern does not need image registration. There are ten fused normalized patterns corresponding to each of the left and right eye of a subject, so in total, we obtain 800 fused normalized patterns from multispectral database.

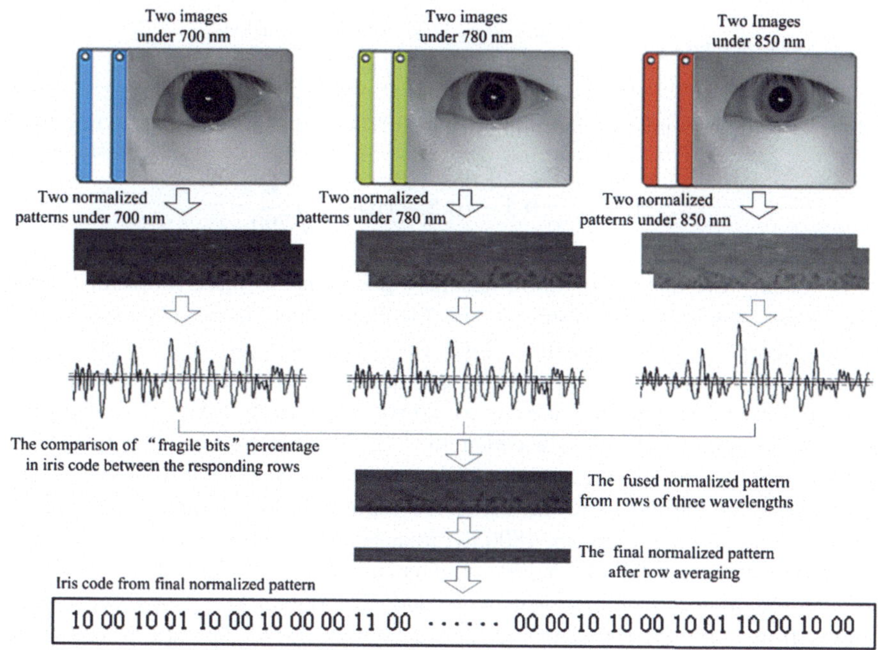

Fig. 3.14 The diagram of multispectral image fusion

The iris codes generated from the rows of normalized pattern under different wavelengths cannot be used for matching directly, so we average neighboring rows to create a smaller normalized pattern. In our work, we average three rows together, the pixels in each column over rows No. 1, 2, and 3, and so on. After row averaging, the resolution of final normalized pattern is 360 × 20, which is used when regenerating the new iris code for matching. The diagram of multispectral image fusion is shown in Fig. 3.14.

Row averaging will lead to an improvement in recognition of multispectral fused images, because of the subtle variations that occur in pupil dilation and eye gesture across different wavelengths. For highly dilated pupils and rotating eyes, the higher radial resolution introduces less information but more interference to recognition. By starting with the larger normalized pattern and averaging rows, we are able to preserve as much iris texture information as possible for multispectral image fusion, while minimizing the amount of duplicate texture information generated from highly dilated pupils and deleting the texture misalignment caused by the eyes with different rotation.

Inspired by the matching scheme of Daugman (2007), the binary Hamming distance is used and the similarity between two iris images is calculated by using the exclusive OR operation. Based on two irides whose two phase code bit vectors are denoted as {codeA, codeB} and mask bit vectors denoted as {maskA, maskB}, we can compute the raw Hamming distance HDraw as follows:

$$\mathrm{HD_{raw}} = \frac{||(\mathrm{codeA} \otimes \mathrm{codeB}) \cap \mathrm{maskA} \cap \mathrm{maskB}||}{||\mathrm{maskA} \cap \mathrm{maskB}||}. \tag{3.4}$$

and then get the normalized Hamming distance HDnorm using a rescaling rule like:

$$\mathrm{HD_{norm}} = 0.5 - (0.5 - \mathrm{HD_{raw}})\sqrt{\frac{n}{790}} \tag{3.5}$$

We usually evaluate recognition accuracy according to three indicators: FAR (false acceptance rate, a measure of the likelihood that the access system will wrongly accept an access attempt), FRR (false rejection rate, the percentage of identification instances in which false rejection occur), and EER (equal error rate, the value where FAR and FRR are equal). So we calculate the FAR, FRR, and EER based on the images of single wavelength 700, 780, 850 nm and the images of multispectral fusion. The intra-spectral genuine and intra-spectral impostor scores are used to compute the EER and FRR. We capture 1600 iris images under each wavelength, so in total, there are 4800 iris images across all three wavelengths. For each wavelength, a total of 15,200 intra-spectral genuine scores and 1,264,000 intra-spectral impostor scores were generated. We obtain 800 fused normalized patterns from multispectral database, and a total of 3,600 intra-spectral genuine scores and 316,000 intra-spectral impostor scores were generated. The normalized histogram plots of $\mathrm{HD_{norm}}$ for the images of single wavelength and the images of multispectral fusion are shown in Fig. 3.15.

In Fig. 3.15, each of four images is composed of two parts: The lower is the matching score distribution, and the upper is the magnified distribution near the cross-point of genuine curve (the left blue one) and impostor curve (the right red one). The blue curve which is to the right of the cross-point means FRR, while the red curve which is to the left of the cross-point means FAR. a–d show the same characteristics: The intra-spectral genuine scores of different wavelengths have similar median values and models and are mostly spread around the corresponding median value. The intra-spectral impostor scores are observed to be fairly well separated from the intra-spectral genuine scores.

The EERs (where FRR = FAR) and FRRs (where the highest degree of FAR accuracy can be obtained by the user, FAR = 0) based on the images of single wavelength and the images of multispectral fusion can be clearly compared. The EER (where FRR = FAR) and FRR (where FAR = 0) reach their lowest values based on the images of multispectral fusion, while the ones based on the images of single wavelength are relatively higher, which means the best recognition performance is achieved by multispectral fusion, see Table 3.4. According to the FRR (where FAR = 0), 1-D Log-Gabor wavelet recognition method with multispectral image fusion is 84 % lower than the one under 850 nm without multispectral image fusion, 68 % lower than the one under 780 nm without multispectral image fusion, and 54 % lower than the one under 700 nm without multispectral image fusion. According to the EER (where FRR = FAR), 1-D Log-Gabor wavelet recognition

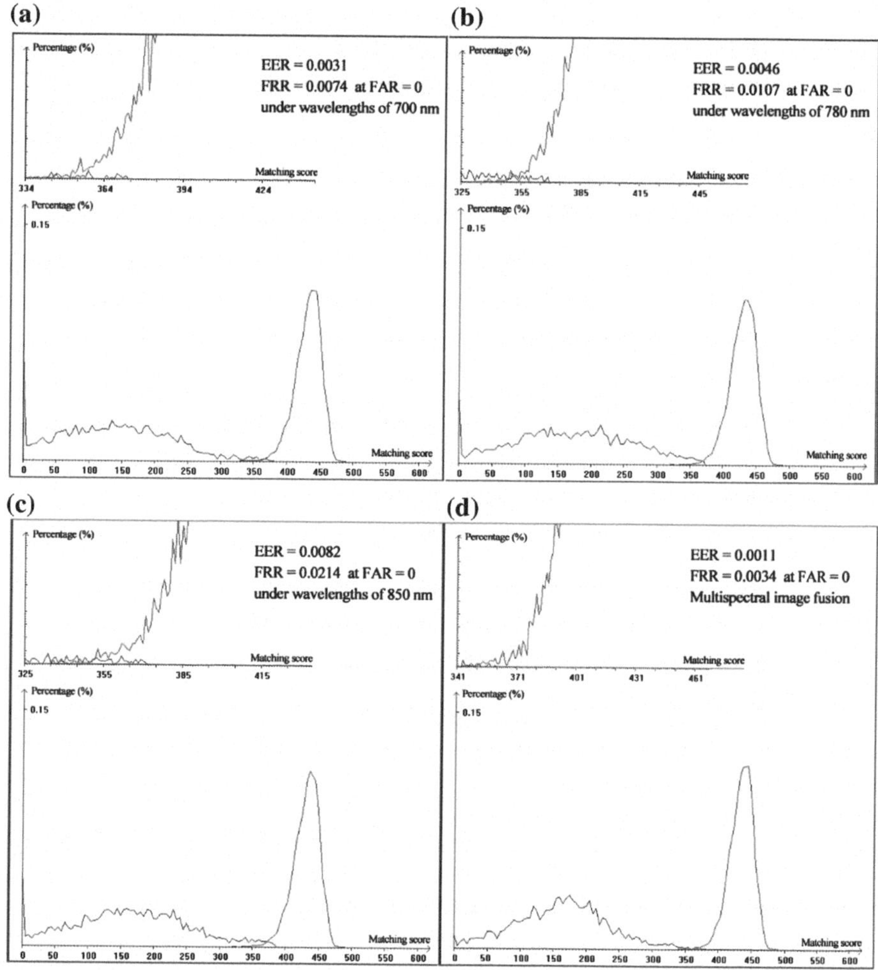

Fig. 3.15 EER and FRR (FAR = 0) based on the images of three wavelengths and the images of multispectral fusion. **a** 700, **b** 780 nm, **c** 850 nm, **d** multispectral fusion

Table 3.4 The comparison of EER and FRR

	700 nm	780 nm	850 nm	Multispectral fusion
FRR at FAR = 0	0.0074	0.0107	0.0214	0.0034
EER	0.0031	0.0046	0.0082	0.0011

method with multispectral image fusion is 86 % lower than the one under 850 nm without multispectral image fusion, 76 % lower than the one under 780 nm without multispectral image fusion, and 64 % lower than the one under 700 nm without multispectral image fusion. From the higher recognition accuracy based on the images of multispectral fusion, we can get three inferences, which are all innovative: First, the multispectral iris images captured by the proposed system are good enough for image fusion and recognition, so the proposed multispectral iris image capture device meets the design requirements and can be used for the rapid iris acquisition within 2 or 3 s, which is much less than the time other similar device takes. Second, the proposed multispectral image-level fusion method is effective and can achieve higher recognition accuracy than the traditional method. Third, the integrated multispectral iris recognition system consisting of the image acquisition device and the recognition server is feasible, and can complete a full process of multispectral online iris identification, which is the first successful attempt of practical application of the iris recognition with multispectral image fusion.

Experiments were also carried out to investigate the cause of a few large intra-spectral genuine scores, which means the multispectral fused images from the same eye sometimes are not very similar. This can be explained with two reasons: First, some interference factors, such as eyelids and eyelashes, maybe occlude the effective regions of the iris, influencing the results of comparison in the "fragile bits" percentage. If more accurate occlusion detection algorithm is used, the performance of multispectral recognition will be better.

Second, and most importantly, some obvious variations occur in pupil dilation and eye gesture across different wavelengths in one data collection cycle, as shown in Fig. 3.16. It is a low probability, but certainly possible, resulting in the multispectral fused image with some degree of texture misalignment. This problem cannot be totally solved by image registration, because no obvious feature in iris area can be used as the reference point. There are two effective solutions for this problem: First, develop higher speed multispectral iris capture system, whose capture cycle is much less than the physiological response cycle of eyes, so we can use iris images without any obvious variations occurring in pupil dilation and eye gesture for image fusion. This is an upgraded version of the proposed capture device, using much higher frame rate CCD camera and faster focusing speed lens, based on the same design ideas and hardware architecture. Second, develop the contemporaneous capture device which can collect multispectral iris images

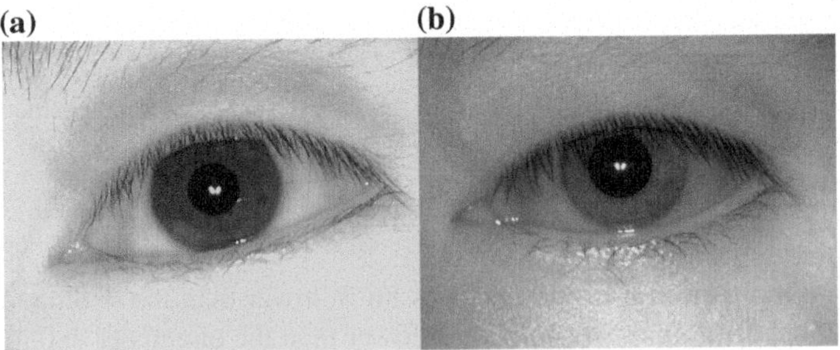

Fig. 3.16 Two images with obvious variation in dilation and rotation captured in one multispectral data collection cycle, **a** under 700 nm, **b** under 850 nm

simultaneously, and all iris images across different wavelengths are captured from the same eye at the same time, ensuring the image registration in the capture process. This is another new attempt based on totally different design ideas and hardware architecture. These two kinds of capture system are all the next focus of our research.

From the above recognition results, we can conclude that the multispectral iris images captured by our proposed device are good enough for iris fusion and recognition.

3.4 Summary

A high-speed multispectral iris capture system has been proposed. Using this system, a complete capture cycle (including three wavelengths) can be completed within 2 or 3 s, much faster than other multispectral iris capture devices. The system consists of the following four parts: (1) capture unit; (2) illumination unit; (3) interaction unit; and (4) control unit. It uses a Sony CCD camera and an automatic focusing lens as the capture unit, and the working distance is about 300 mm. Two groups of matrix-arrayed LEDs across three different wavelengths (including visible light and near-infrared light) are used as the main multispectral illumination unit, and three LEDs in triangular arrangement are used as the auxiliary illumination unit, specially designed for pupil location. We design an interaction unit including the infrared distance measuring sensor and the speaker to realize the exact focusing range of the lens via the real-time feedback of the voice prompts. The novel design on the control unit synchronizes the previous three units and makes it capturing with high speed, easy to use, and nonintrusive for users.

A series of experiments is performed to evaluate the performance of the proposed multispectral iris capture device. A multispectral iris image database is created by the proposed capture device, and then, we use the iris image-level fusion to further investigate the effectiveness of the proposed capture device by the 1-D Log-Gabor wavelet filter approach. Experimental results have illustrated the encouraging performance of the current design.

In summary, we conclude that our high-speed multispectral iris capture system can achieve good performance at a reasonable price so that it becomes suitable for civilian personal identification applications. For further improvement of the system, we will focus on the following three issues: (1) embedding the light flicker to illumination unit to control the pupil dilation when capturing (Peretto et al. 2007); (2) conducting multispectral fusion and recognition experiments on a large number of iris databases in various environments for the system to be more stable and reliable; (3) presenting a more critical analysis of the accuracy and performance measurement of the proposed system based on large-scale experimental data (Gamassiet al. 2005).

References

Biom Technol (2005) Iris recognition in focus. Today 13(2): 9–11

Burge MJ, Monaco MK (2009) Multispectral iris fusion for enhancement, interoperability, and cross wavelength matching. In: SPIE Defense, Security, and sensing, pp 73341D–73341D

CASIA Iris Image Database (2005) http://www.cbsr.ia.ac.cn/IrisDatabase.htm

Daugman JG (1993) High confidence visual recognition of persons by a test of statistical independence. IEEE Trans Pattern Anal Mach Intell 15(11):1148–1161

Daugman J (2004) How iris recognition works. IEEE Trans Circ Syst Video Technol 14(1):21–30

Daugman J (2007) New methods in iris recognition. IEEE Trans Syst Man Cybern B Cybern 37(5):1167–1175

Ebisawa Y (1998) Improved video-based eye-gaze detection method. IEEE Trans Instrum Meas 47(4):948–955

Gamassi M, Lazzaroni M, Misino M, Piuri V, Sana D, Scotti F (2005) Quality assessment of biometric systems: a comprehensive perspective based on accuracy and performance measurement. IEEE Trans Instrum Meas 54(4):1489–1496

He X, Yan J, Chen G, Shi P (2008) Contactless autofeedback iris capture design. IEEE Trans Instrum Meas 57(7):1369–1375

Hollingsworth KP, Bowyer KW, Flynn PJ (2009) The best bits in an iris code. IEEE Trans Pattern Anal Mach Intell 31(6):964–973

Iris Recognition Camera System (2009) http://catalog2.panasonic.com/webapp/wcs/stores/servlet/ModelList?storeId=11201&catalogId=13051&catGroupId=21552&surfModel=BM-ET330

Iris Recognition from IRIS ID (2014) http://www.irisid.com/home

Kang BJ, Park KR (2007) Real-time image restoration for iris recognition systems. IEEE Trans Syst Man Cybern B Cybern 37(6):1555–1566

Masek L (2003) Recognition of human iris patterns for biometric identification. Doctoral dissertation, Master's thesis, University of Western Australia

Mobile Dual Iris Capture Device (2014) http://www.crossmatch.com/i-scan-2/

Ngo HT, Ives RW, Matey JR, Dormo J, Rhoads M, Choi D (2009) Design and implementation of a multispectral iris capture system. In: Signals systems and computers 2009 conference record of the forty-third Asilomar conference, pp 380–384

Oki Introduces the IRISPASS®-WG Iris Recognition System with Automatic Iris Scanning Function (2002) http://www.oki.com/en/press/2002/z02011e.html

Park KR, Kim J (2005) A real-time focusing algorithm for iris recognition camera. Syst Man Cybern Part C: IEEE Trans Appl Rev 35(3):441–444

Peretto L, Rovati L, Salvatori G, Tinarelli R, Emanuel AE (2007) A measurement system for the analysis of the response of the human eye to the light flicker. IEEE Trans Instrum Meas 56 (4):1384–1390

Peters TH (2009) Effects of segmentation routine and acquisition environment on iris recognition. Doctoral dissertation, University of Notre Dame

Ross A, Pasula R, Hornak L (2006) Exploring multispectral iris recognition beyond 900 nm. In: Proceedings of the 2006 conference on computer vision and pattern recognition workshop: 51

Shi P, Xing L, Gong Y (2003) A quality evaluation method of iris recognition system. Chin Pat 1 (474): 345

Tan T, Zhu Y, Wang Y (1999) Iris image capture device. Chin Pat 2(392):219

Vilaseca M, Mercadal R, Pujol J, Arjona M, de Lasarte M, Huertas R, Imai FH (2008) Characterization of the human iris spectral reflectance with a multispectral imaging system. Appl Opt 47(30):5622–5630

Wildes RP (1997) Iris recognition: an emerging biometric technology. Proc IEEE 85(9):1348–1363

Wilkerson CL, Syed NA, Fisher MR, Robinson NL, Albert DM (1996) Melanocytes and iris color: light microscopic findings. Arch Ophthalmol 114(4):437–442

Chapter 4
Feature Band Selection for Multispectral Iris Recognition

Abstract This work uses East Asian irides as research subjects and explores the possibility of clustering spectral wavelengths based on the maximum dissimilarity of iris textures. The eventual goal is to determine how many bands of spectral wavelengths will be enough for black-based iris multispectral fusion, and find these bands, which will provide an important standard for selecting bands of spectral wavelengths for iris multispectral fusion, especially for the black iris recognition of East Asians. A multispectral acquisition system is first designed for imaging the iris at narrow spectral bands in the range of 420–940 nm. Next, a set of 60 human black iris images which correspond to the right and left eyes of 30 different subjects are acquired for an analysis. Finally, we have determined that 3 clusters are enough to represent the 10 feature bands of spectral wavelengths from 545 to 940 nm, using the agglomerative clustering based on an improved multigroup two-dimensional principal component analysis [$(2D)^2$PCA]. The experimental results suggest: (a) the number, center, and composition of clusters of spectral wavelengths and (b) the interference and potential impact of interference on the performance of iris multispectral fusion.

Keywords Mutlispectral iris · $(2D)^2$PCA · Feature band · Clustering

4.1 Introduction

Traditionally, only a narrow band of the near-infrared (NIR)[1] spectrum (750–850 nm) was utilized for iris recognition systems since this will alleviate any physical discomfort from illumination, reduce specular reflections, and increase the amount of iris texture information captured for some of the iris colors. Commercial iris recognition systems predominately operate in the NIR range of the electromagnetic spectrum. The spectrums indicate that current systems are using wavelengths that peak around 850 nm (Panasonic and Oki), with a narrow band pass.

[1] http://en.wikipedia.org/wiki/Near_Infrared.

D. Zhang et al., *Multispectral Biometrics*,
DOI 10.1007/978-3-319-22485-5_4

However, some systems traverse into the range of 750 nm (LG) and use multiple wavelength illumination to image the iris. The infrared light is invisible to the human eye, and the intricate textural pattern represented in different colored irides is revealed under an NIR range of illumination. The texture of the iris in IR illumination has been traditionally used as a biometric indicator (Boyce 2006).

However, the textural content of the iris has complex components, including numerous structures and various pigments, both fibrous and cellular, which are contained on the anterior surface, including ligaments, crypts, furrows, collarettes, moles, and freckles. The NIR wavelengths can penetrate melanin, showing a texture which cannot be easily observed in the visible spectrum, but the cost is substantially high. Most of the texture presented in the NIR spectrum is only generated by the iris structures, not by the pigments. The effect of melanin, the major color-inducing compound, is negligible on the NIR wavelengths in iris recognition. But, melanin is always imaged in certain wavelength for extraction and classification, such as in tongue image processing (Liu et al. 2007).

The above study has inspired us into considering that the iris textures generated outside the NIR spectrum may have more information over those that are only generated in the NIR spectrum, because melanin can be present in the shorter wavelengths and becomes another major source of iris texture.

Previous research has shown that matching performance is not invariant to iris color and can be improved by imaging outside the NIR spectrum and that the physiological properties of the iris (e.g., the amount and distribution of melanin) impact the transmission, absorbance, and reflectance of different portions of the electromagnetic spectrum and the ability to image well-defined iris textures (Wilkerson et al. 1996). Performing multispectral fusion at the score level is proven feasible (Ross et al. 2006), and the multispectral information is used to determine the authenticity of the imaged iris (Park and Kang 2007).

When we are ready to accept that multispectral iris fusion can improve matching performance, which has been proven by several previous experiments, two questions need to be answered as the cornerstones that support the above conclusion. Question 1: how do the colors of the irides influence the matching performance of multispectral iris recognition? Question 2: how does the iris texture generated from the structures and pigments change with the illumination of different spectral wavelengths?

Some researchers have tried to answer Question 1 and achieved some initial results. Burge and Monaco (2009) demonstrated that iris texture increases with the frequency of illumination for lighter colored sections of the iris and decreases for darker sections. This means that the effects of an illumination wavelength on various colored sections of the iris are not the same; sometimes, the texture increases, and sometimes, it decreases, depending on the color of the iris. Hence, the feasibility of multispectral iris recognition cannot be explored from only the perspective of an electromagnetic spectrum, and the colors of the irides should be studied as a very important factor, combined with illumination wavelengths.

Although these previous studies did not really answer Question 1, at least they clarified a basic principle: The accuracy of conclusions is based on the basis that the iris images used in a multispectral study should belong to a certain color classification. This is because the same research methods may produce an entirely different conclusion on the iris images of different colors.

The irides of East Asians are a good object of study. Franssen et al. (2008) independently ranked 24 photographs from least (number 1) to most (number 24) average iris pigmentation. The iris color distribution of East Asians is black based and concentrated. The following are characteristics of East Asian irides: dark brown or black in color, greater amount of melanin than general irides, and a melanin distribution that covers the entire iris texture region.

In terms of Question 2, there is no related in-depth research and no published results. From previous studies, we are only able to observe the phenomenon in which iris images captured across multispectral wavelengths show differences in the amount and distribution of texture and should be used for feature fusion in order to increase the diversity of iris textures, but we do not know the specific mechanisms of the iris texture that change with multispectral wavelengths. Studying the above mechanism has great significance, especially in the choice of which band of the electromagnetic spectrum can be used for iris fusion.

As usually seen in the iris images captured in multispectral wavelengths, iris textures generated from the structures and pigments may mix or overlap with each other. Hence, if the bands of the electromagnetic spectrum are not suitable, only an average spectral reflectance profile or equivalently a mean color from the mixture structure and melanin, usually the black color, can be extracted from the collected data.

Since we lack understanding of the mechanisms of iris texture that change with multispectral wavelengths, we can only try various combinations of iris images across all spectral wavelengths randomly and choose the best fusion strategy based on the comparisons of different matching performances. The disadvantage is obvious; the number of spectral wavelength bands used for fusion is difficult to determine.

On the contrary, if we can cluster the spectral wavelengths based on the maximum dissimilarity of the corresponding iris texture and choose an iris image from each classification of spectral wavelength for fusion, the two conditions of fusion strategy—completeness and no redundancy—will be simultaneously met, and the best fusion result will most likely be achieved.

This work uses East Asian irides as research subjects and explores the possibility of clustering spectral wavelengths from 420 to 940 nm based on the maximum dissimilarity of iris textures captured in the corresponding spectral wavelengths, which will provide an important standard for selecting bands of the spectral wavelengths for iris multispectral fusion. The eventual goal is to determine how many bands will be enough for black-based iris multispectral fusion and find these bands. This research represents the first attempt in the literature to investigate the irides of East Asians in a multispectrum analysis.

4.2 Data Collection

4.2.1 Overall Design

In this work, we have analyzed the feasibility of a conventional multispectral system based on a charge-coupled device (CCD) monochrome camera to capture iris images. One of the most challenging aspects of this research is the design of a multispectral image acquisition system.

The multispectral system developed consists of a CCD camera with a Sony ICX205 HAD CCD sensor (spectral response ranges between 400 and 1000 nm with a peak at 550 nm, 1.4 megapixels 1360 × 1024), macro-manual focus (MF) lens, illumination system, and Meadowlark selectable bandwidth tunable optical filter TOF-SB-VIS (see Fig. 4.1).

The optical path is as follows: First, the broadband light from the illumination system is delivered to the eye, and then, the reflected light from the subject's eye is

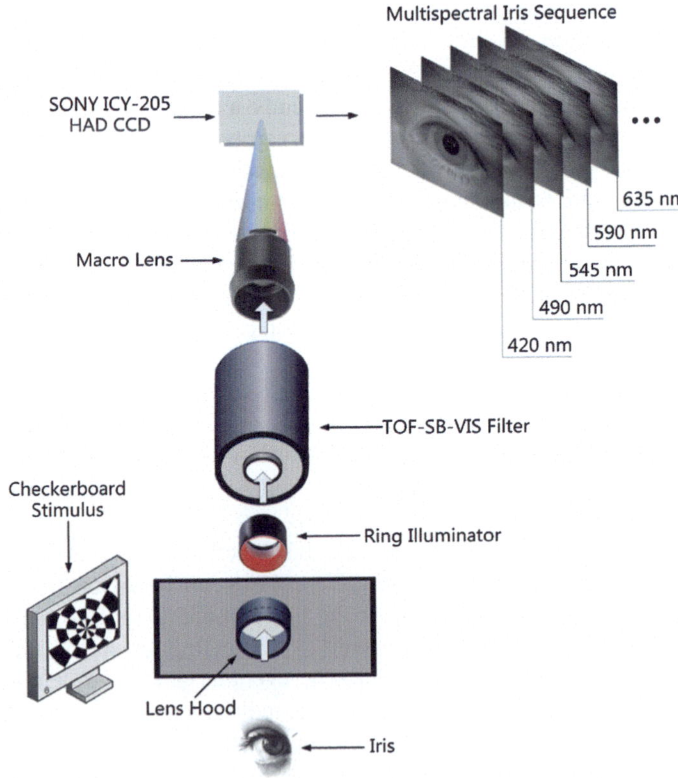

Fig. 4.1 The structure of the multispectral iris acquisition system

collected through the center of the ring illuminator, through the band-pass filter, TOF-SB-VIS, and imaged by the ICX205 camera using a macro lens.

The camera with the Sony ICX205 HAD CCD sensor has exceptional features, including high resolution, high sensitivity, and low dark current, which are all important to multispectral imaging. The spectral response from the 400 to 1000 nm wavelength (the short-wavelength infrared (SWIR) band) is not very uniform, but we verified that the CCD response does not introduce significant errors into the experimental values after the optimization of the multispectral system.

TOF-SB-VIS is a new tunable optical filter with user selectable bandwidths and a variable full width at half maximum (FWHM) available through Meadowlark Optics. By utilizing multiple liquid crystal variable retarders and polarizers, this tunable filter allows the user to switch between any wavelengths from 420 to 1100 nm. In this research, the band-pass wavelengths of the TOF-SB-VIS are switched at 420, 490, 545, 590, 635, 665, 700, 730, 780, 810, 850, and 940 nm, which correspond to 12 kinds of narrow band LEDs that are used in sequence to image a subject's eye across the visible and NIR bands.

The illumination system has two parts: the multispectral light source and the checkerboard stimulus, which are especially designed to meet the special needs of iris image acquisition. The multispectral light source is a ring illuminator with six narrow wavelength band LED lamps, which is located in front of the TOF-SB-VIS, between the imaging device and the subject, and can be manually switched, allowing illumination of the captured iris with a 90° angle. In accordance with the spectral range of interest, narrow band LEDs are selected at 12 wavelengths, so that the corresponding band-pass wavelength outputs are delivered to the eye.

4.2.2 Checkerboard Stimulus

The checkerboard stimulus that is composed of an LCD screen that plays a circular checkerboard reversal pattern and a supporting structure allows illumination of the captured iris with a 75° angle of incidence and obtains a rather uniform luminous field on the eye. The checkerboard reversal pattern on the screen can flicker between black on white and white on black (invert contrast) at a certain frequency without change of space-averaged luminance of the eye (see Fig. 4.2a).

Sun et al. (1998) investigated human pupillary responses evoked by visual spatial patterns and demonstrated that pupillary constriction can be induced not only by increment of luminance, but also by change of gratings or checkerboards without change of space-averaged luminance. In their experiment, the checkerboard was used as the standard stimulus on the pupil. This explains the role of checkerboard stimulus in image capturing; it causes pupillary constriction, which can be an effective solution to the problem of the changing radius of the pupil.

The radius of the pupil constantly changes because of pupil dilation and contraction, an involuntary physiological mechanism. As a result, the pupil is constantly compressing or expanding the biological tissue that gives structure to the iris. This

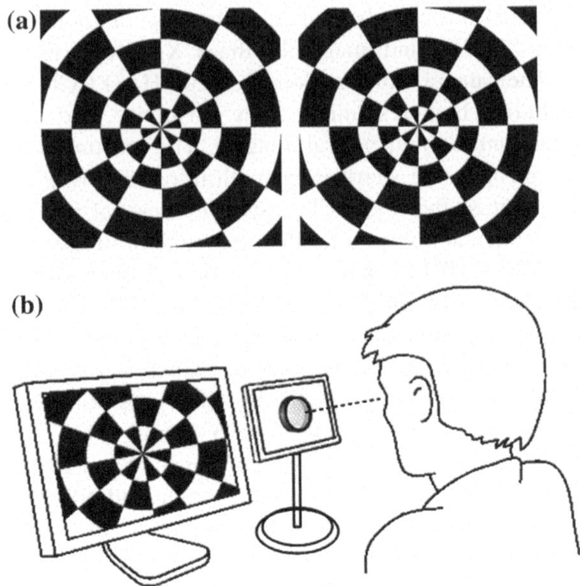

Fig. 4.2 **a** The circular checkerboard reversal pattern that stimulates the constriction of pupils. **b** Iris image acquisition

motion has several consequences: First, it generates lateral pressure on the tissue that may cause small parts of the tissue to fold underneath other parts or become newly visible. This means that small parts of the pattern structure will disappear and will not be recovered by any model of in-plane deformation. As a result, no technique can correct this phenomenon, which is lethal and unacceptable for any spectral clustering method based on an iris texture analysis.

The second consequence of pupil motion is a remapping of the iris pattern. Daugman (1993) presented a method that represents an iris pattern in a polar coordinate system. In general, the alteration from pupil motion can be described using a one-to-one coordinate mapping function. If the motion is perfectly linear along the radial direction, the segmentation process described by Daugman would normalize for this change. However, in real iris pattern observations, the motion is approximately linear at best and not necessarily limited to the radial direction. This more complex motion leads to minor relative deformations in normalized patterns (Thornton et al. 2007).

Hollingsworth et al. (2009) studied the effect of texture deformations caused by pupil dilation on the accuracy of iris biometrics and found that when matching two iris images (enrollment and recognition) of the same person, larger differences in pupil dilation yield higher template dissimilarities and therefore a greater chance of a false non-match. The above research demonstrated that texture nonlinear deformations caused by pupil dilation are sometimes so serious as to let iris recognition algorithms make this wrong decision: Two iris images of the same person are considered to originate from two different persons. Hence, if we ignore the problems associated with variations in pupil dilation among the iris images captured in multispectral wavelengths, some significant errors will be introduced into clustering spectral wavelengths based on the iris texture analysis.

In this work, the basic idea for solving the problem of texture deformations caused by pupil dilation is as follows: Multispectral iris images are used with a similar degree of pupil dilation as the research subjects to minimize the interference caused by pupil dilation. Taking into account the amount of iris texture information, images of pupillary constriction will be more suitable for the requirements of this experiment, so the acquisition system should be designed to capture such images. Consequently, this warrants the need for a checkerboard stimulus.

In fact, in the experiment, we also tried another method to cause pupillary constriction. We let the multispectral light source flash according to a certain frequency during image acquisition, which can produce a stimulating effect similar to the checkerboard stimulus. However, the flashing multispectral light source has two weaknesses: First, the human eye is not sensitive to the bands of the NIR spectrum (more than 700 nm) (NIR), so the flashing light source of the NIR is unable to cause pupillary constriction. Second, the flashing light source changes the space-averaged luminance and results in an inconsistent and unstable level of iris exposure in digital images (unless there is synchronization between flash and capture, but at the cost of increasing system complexity and reducing acquisition speed), which will introduce new interference to the texture analysis. So, the checkerboard stimulus is a better solution than the flashing light source.

4.2.3 Data Collection

A set of 60 human irides which correspond to the right and left eyes of 30 different subjects was captured by a multispectral acquisition system. The 60 samples covered a wide range of East Asian iris colorations and structures, such as dark brown and black, and rich and sparse textures.

Twenty images were taken at a specific interval of time, which was 200 ms apart, from each the left and the right eye of a subject under 420, 490, 545, 590, 635, 665, 700, 730, 780, 810, 850, and 940 nm wavelengths, respectively. During the image capture, the subject watched the center of a ring illuminator and the checkerboard reversal pattern on the screen flickered between black on white and white on black (invert contrast) at a frequency of 2 times per second (see Fig. 3.2b).

According to previous research (Li and Sun 2005), a fast pupillary constriction can last about 500 ms before returning to the initial level. Hence, there are at least 3 images that cover the time window of 500 ms, recording a whole process of pupillary contractions. According to this proportion, the capturing of more than 20 images, associated with a single spectral wavelength, will last more than 4 s and record several pupillary contractions. Sample multispectral iris images which pertain to a single eye are shown in Fig. 4.3. We deliberately selected images whose pupil radius has a relatively consistent and small size, to observe the differences in iris texture across all spectral wavelengths.

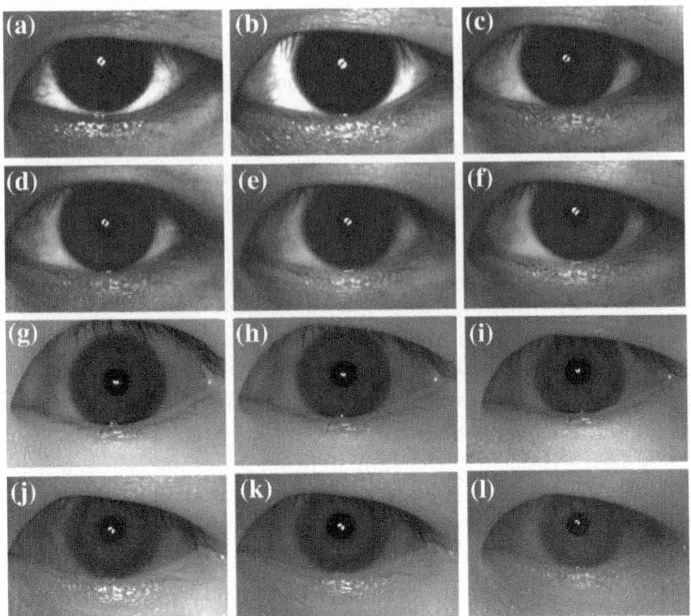

Fig. 4.3 Sample images obtained at wavelengths of **a** 420 nm, **b** 490 nm, **c** 545 nm, **d** 590 nm, **e** 635 nm, **f** 665 nm, **g** 700 nm, **h** 730 nm, **i** 780 nm, **j** 810 nm, **k** 850 nm, and **l** 940 nm

4.3 Feature Band Selection

4.3.1 Data Organization of Dissimilarity Matrix

Suppose we have **B** bands of spectral wavelengths and S iris images from the same eye of the same subject captured under each band of spectral wavelengths, so the total number of iris images is **S** × **B**. Based on the definition of the distance, we can compute the distance between any pair of images, T_i and T_j (whether from the same or different spectral wavelength) as the generic dissimilarity measure. We embed the resulting dissimilarity data in Matrix A ($N \times N$), and the elements of A are defined as:

$$A(i,j) = d(T_i, T_j), \quad i,j = 1, \ldots, N, \;\; N = S \times B \tag{4.1}$$

The image axis runs on the horizontal (left to right) and vertical (top to bottom) axes of **A** and along its main diagonal, where self-similarity is maximal. According to the order of time stamp and increasing wavelength, N iris images of the same eye can be arranged as a sequence along the axis. The iris images of the same spectral wavelength are all continuously captured, so these images will be adjacent in the axis (see Fig. 4.4).

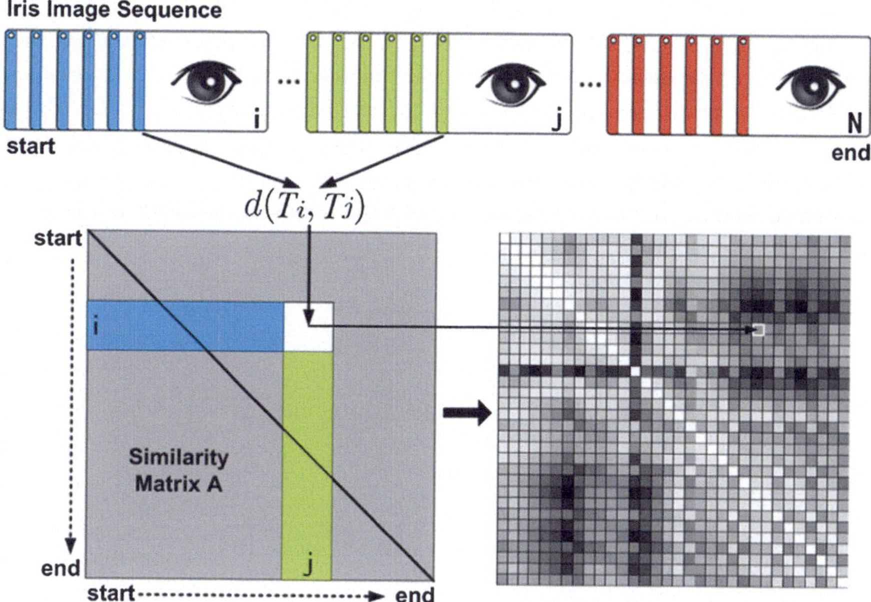

Fig. 4.4 The structure of the image-indexed dissimilarity Matrix **A**

The image-indexed dissimilarity Matrix **A** can be visualized as a square image. Each pixel i, j is colored with a gray scale value proportional to the dissimilarity measure $d(T_i, T_j)$. In these visualizations, we can clearly see the structure of a multispectral iris image sequence. Regions of the highest similarity, generated from a similar iris texture within the same spectral wavelength, appear as the brightest squares on the diagonal. The relatively brighter rectangular regions off the main diagonal indicate the similarity between the different spectral wavelengths (see Fig. 4.4).

4.3.2 Improved $(2D)^2$PCA

Now, we will determine the specific definition of the distance measure, and the $(2D)^2$PCA method (Zuo et al. 2006) is a good choice for iris images. The $(2D)^2$PCA method can alleviate the small sample size problem in a subspace analysis and preserve well the image local structural information.

In this work, the $(2D)^2$PCA analysis is simultaneously done for different bands of spectral wavelengths. An improved $(2D)^2$PCA will be presented which allows us to analyze group elements that have common principle components (PCs). From a statistical point of view, simultaneously estimating PCs in different bands will result in a joint dimension reducing transformation. This multigroup $(2D)^2$PCA, the so-called common (2D)^2 PC analysis, yields a joint eigenstructure across two bands of spectral wavelengths, which will be used in the distance measurement between two iris images captured under different spectral wavelengths.

We denote by X_s^b, the sth iris image of the band bth, and by $X_s^{b+b'}$, the sth iris image of bands bth and b'th (including $2S$ iris images). $X_s^{b+b'}$ is an $I_r \times I_c$ matrix, where I_r and I_c represent the numbers of rows and columns of the image. The covariance matrices along the row and column directions of the bands bth and b'th are computed as follows:

$$\begin{aligned} G_1^{b+b'} &= \frac{1}{2S}\sum_{s=1}^{2S}\left(X_s^{b+b'} - \overline{X^{b+b'}}\right)^T\left(X_s^{b+b'} - \overline{X^{b+b'}}\right) \\ G_2^{b+b'} &= \frac{1}{2S}\sum_{s=1}^{2S}\left(X_s^{b+b'} - \overline{X^{b+b'}}\right)\left(X_s^{b+b'} - \overline{X^{b+b'}}\right)^T \end{aligned} \tag{4.2}$$

where $\overline{X^{b+b'}} = \frac{1}{2S}\sum_{s=1}^{2S} X_s^{b+b'}$.

The project matrix $V_1^{b+b'} = \left[V_{11}^{b+b'}, V_{12}^{b+b'}, \ldots, V_{1k_1^{b+b'}}^{b+b'}\right]$ is composed of the orthogonal eigenvectors of $G_1^{b+b'}$ which correspond to $k_1^{b+b'}$ the largest eigenvalues, and the projection matrix $V_2^{b+b'} = \left[V_{21}^{b+b'}, V_{22}^{b+b'}, \ldots, V_{2k_2^{b+b'}}^{b+b'}\right]$ consists of the orthogonal eigenvectors of $G_2^{b+b'}$ which correspond to the $k_2^{b+b'}$ largest eigenvalues. $k_1^{b+b'}$ and $k_2^{b+b'}$ can be determined by a threshold:

$$\frac{\sum_{j_c=1}^{k_1^{b+b'}} \lambda_{1j_c}^{b+b'}}{\sum_{j_c=1}^{I_c} \lambda_{1j_c}^{b+b'}} \geq C_u \tag{4.3}$$

$$\frac{\sum_{j_r=1}^{k_2^{b+b'}} \lambda_{2j_r}^{b+b'}}{\sum_{j_r=1}^{I_r} \lambda_{2j_r}^{b+b'}} \geq C_u \tag{4.4}$$

where $\lambda_{11}^{b+b'}, \lambda_{12}^{b+b'}, \ldots, \lambda_{1I_c}^{b+b'}$ are the first I_c largest eigenvalues of $G_1^{b+b'}$, $\lambda_{21}^{b+b'}, \lambda_{22}^{b+b'}, \ldots, \lambda_{2I_r}^{b+b'}$ are the first I_r largest eigenvalues of $G_2^{b+b'}$, and C_u is a preset threshold. Now, we yield the joint eigenstructure $V_1^{b+b'}$ and $V_2^{b+b'}$ across two bands of spectral wavelengths bth and b'th, and the space spanned by these eigenvectors is specified for the distance measurement of iris images from these two corresponding bands (see Fig. 4.5).

An image T^b of the band bth will be matched with another image $T^{b'}$ of the band b'th, and the two images T^b and $T^{b'}$ are all captured from the same eye of the same subject. These two images should be projected to $\hat{T}^b$ and $\hat{T}^{b'}$ by $V_1^{b+b'}$ and $V_2^{b+b'}$, and the distance of T^b and $T^{b'}$ is defined as:

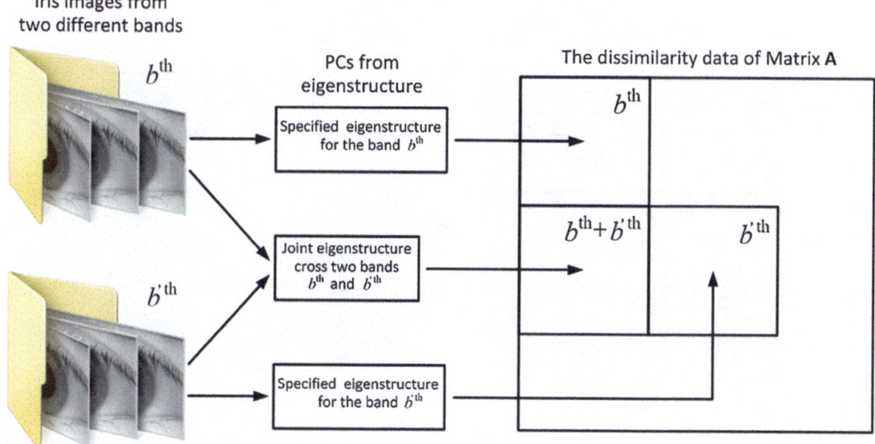

Fig. 4.5 Two types of principle components generated from specified and joint eigenstructures, and the correspondence with the dissimilarity data of Matrix **A**

$$d^{b+b'} = \left\| \hat{T}^b - \hat{T}^{b'} \right\| = \left\| V_2^{b+b'T} T^b V_1^{b+b'} - V_2^{b+b'T} T^{b'} V_1^{b+b'} \right\| \tag{4.5}$$

In terms of the distance of the two images within band bth, the above algorithm will be slightly simplified. The covariance matrices along the row and column directions of band bth are computed as:

$$\begin{aligned} G_1^b &= \frac{1}{S}\sum_{s=1}^{S}\left(X_s^b - \overline{X^b}\right)^T\left(X_s^b - \overline{X^b}\right) \\ G_2^b &= \frac{1}{S}\sum_{s=1}^{S}\left(X_s^b - \overline{X^b}\right)\left(X_s^b - \overline{X^b}\right)^T \end{aligned} \tag{4.6}$$

where $\overline{X^b} = \frac{1}{S}\sum_{s=1}^{S} X_s^b$, and other calculations are similar to the above algorithm. Finally, we get the eigenstructure V_1^b and V_2^b which corresponds to band bth, and the space spanned by these eigenvectors is specified for the distance measurement of iris images from the same band (see Fig. 4.5).

4.3.3 *Low-Quality Evaluation*

Taking into account the possibility of some low-quality iris images in sequence, such as motion blurs, camera noise, or inaccurate focusing, which will introduce interference to the dissimilarity data of Matrix **A**, we need a statistical measure to estimate image quality. Generally, when low-quality image is matched with other

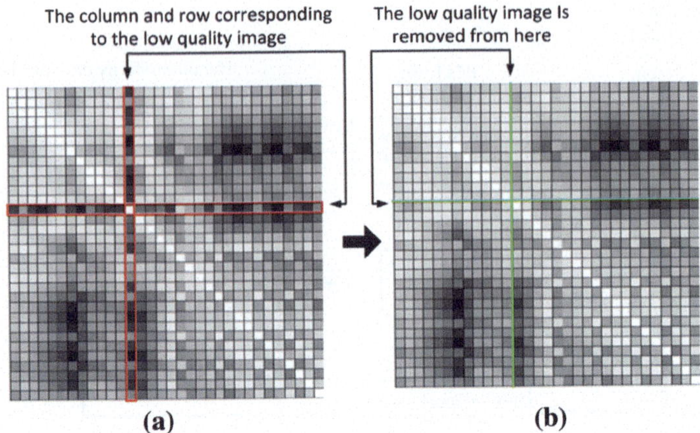

Fig. 4.6 **a** The image-indexed dissimilarity Matrix A with all images, and the column and row that correspond to the low-quality image is marked with *red lines*. **b** After the evaluation and removal of low-quality images, Matrix A without the corresponding column and row, and the original position of the low-quality image is marked with *green lines* (color figure online)

images, the distance between them will be relatively large, so we can compute the sum of elements in each column or row of Matrix **A** as:

$$A_c(i) = \sum_{j=1}^{N} d(T_i, T_j), \quad i = 1, \ldots, N \tag{4.7}$$

$A_c(i)$ is compared with a preset threshold $\boldsymbol{Th}$: If $A_c(i) \geq \boldsymbol{Th}$, the image T_i will be removed from the Matrix **A**; if $A_c(i) < \boldsymbol{Th}$, the image T_i will be retained (see Fig. 4.6a, b).

4.3.4 Agglomerative Clustering Based on the Global Principle

Before clustering, we can search for a pair of similar bands to merge into a single new segment based on the following assumption, the so-called "**local principle**". There is a continuous gradient of changes with iris texture in images across all spectral wavelengths; thus, some adjacent spectral wavelengths may have more similarity in iris image and can be merged into a single segment first. Note that only the bands adjacent in spectral wavelengths can be chosen for merging; hence, the minimum size of the merge depends on the band and not the image.

Matrix **A** can be considered as the composition of 2 × 2 checkerboards of different scales. Using the cosine metric, similar regions will be close to 1, while dissimilar regions will be closer to −1. Finding the cutoff point of the spectral wavelength is as simple as finding the crux of the checkerboard. This can be done by correlating Matrix **A** with a checkerboard kernel (Foote 2000), and a high value (we call it a novelty score) will be obtained when the two regions are self-similar, but different from each other. Kernels can be smoothed to avoid edge effects using windows that taper toward zero at the edges. Usually, a radially symmetric Gaussian function is used (Baraniuk and Jones 1993). Note that the scale of the checkerboard kernel will be related to the amount of iris images of a specific spectral wavelength and the total amount of iris images across all spectral wavelengths.

We get a series of novelty scores by correlating Matrix **A** with a Gaussian-based kernel, and the peaks of the novelty scores correspond to large changes in the iris image sequence. These points can serve as cutoff points for segmenting the spectral wavelengths and be found where the novelty score exceeds a preset threshold. The cutoff points divide the iris sequence into a set of segments. Figure 4.7 (a) is the image-indexed dissimilarity Matrix **A**, (b) is the novelty scores generated by using a Gaussian-based kernel, and (c) is Matrix **A** separated by the cutoff points that correspond to the three highest novelty scores, which has been marked with arrows. Based on the cutoff points for segmentation, we can calculate a dissimilarity matrix of substantially lower dimension, indexed by the number of segments instead of the number of images.

After the initial segments, there will be a variable amount of images between the different segments, so we can use the Kullback–Leibler (KL) distance as the statistical measure to estimate the dissimilarity between two sets with different data size.

Suppose we have $\boldsymbol{B}$ bands of spectral wavelengths and K segments $\{p_1, \ldots, p_K\}$ of variable lengths ($K \leq \boldsymbol{B}$), and we can compute the mean and covariance for the spectral data in each segment. Inter-segment dissimilarity is defined as the KL distance (Cover and Thomas 1991) between the normal probability densities with the statistics of the segments. The KL distance between the n-dimensional normal

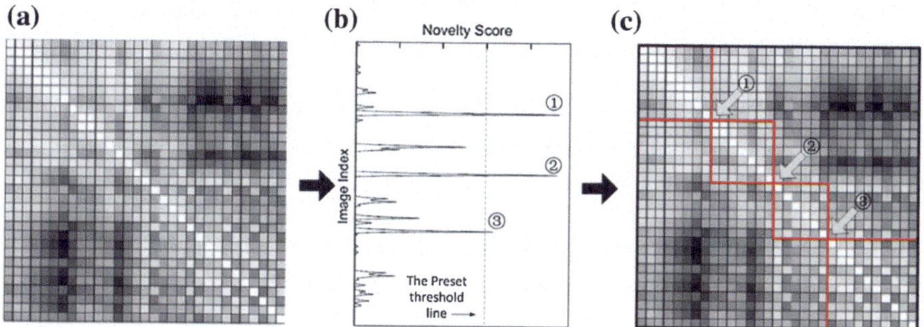

Fig. 4.7 **a** Image-indexed dissimilarity Matrix A. **b** Three highest novelty scores that exceed the preset threshold, and **c** Matrix A separated by the cutoff points that correspond to the highest novelty scores

probability densities $N(\mu_i, \Sigma_i)$ and $N(\mu_j, \Sigma_j)$ is as follows (Ghosal et al. 1999; Ghosal and Van Der Vaart 2001, 2007; Tokdar 2006):

$$d_{\mathrm{KL}}(N(\mu_i, \Sigma_i)\|N(\mu_j, \Sigma_j)) = \frac{1}{2}\left[\log\frac{\det \Sigma_j}{\det \Sigma_i} + Tr(\Sigma_j^{-1}\Sigma_i) - 2\mu_j'\Sigma_j^{-1}\mu_i + \mu_j'\Sigma_j^{-1}\mu_j + \mu_i'\Sigma_j^{-1}\mu_i - n\right] \tag{4.8}$$

Tr means the matrix trace. The KL distance is not symmetric, but a symmetric variation can be constructed from the sum of the two KL distances as (Johnson and Sinanovi 2001):

$$\hat{d}_{\mathrm{KL}}(N(\mu_i, \Sigma_i)\|N(\mu_j, \Sigma_j)) = \frac{1}{2}\left[N(\mu_i, \Sigma_i)\|N(\mu_j, \Sigma_j) + N(\mu_j, \Sigma_j)\|N(\mu_i, \Sigma_i)\right] \tag{4.9}$$

So, the dissimilarity between segments p_i and p_j can be defined as follows:

$$d_{\mathrm{seg}}(p_i, p_j) = \exp(-\hat{d}_{KL}(N(\mu_i, \Sigma_i)\|N(\mu_j, \Sigma_j))), \quad d_{\mathrm{seg}}(p_i, p_j) \in (0, 1] \tag{4.10}$$

We compute the inter-segment dissimilarity between each pair of segments and embed them into a segment-indexed dissimilarity Matrix A_s ($K \times K$), analogous to the image-indexed dissimilarity Matrix $\mathbf{A}$ ($N \times N$), as follows:

$$A_s(i,j) = d_{\mathrm{seg}}(p_i, p_j) \quad i,j = 1, \ldots, K \tag{4.11}$$

We use the agglomerative algorithm (Ozawa 1983) to cluster the K segments $\{p_1, \ldots, p_K\}$ based on the segment-indexed dissimilarity Matrix A_s. The input of agglomerative algorithm is the Matrix A_s ($K \times K$), the initial dissimilarity Matrix $A_s^0 = A_s$ At each level t, when two segments are merged into one, the size of the dissimilarity Matrix A_s^t becomes $(K-t) \times (K-t)$. A_s^t follows A_s^{t-1} by (a) deleting the two rows and columns that correspond to the merged clusters and (b) adding a new row and a new column that contain the distance between the newly formed cluster and the old (unaffected at this level) cluster. The distance between the newly formed cluster C_q (the result of merging C_i and C_j) and an old cluster C_s is defined as the KL distance (see Fig. 4.8).

Applying the above process, we can get the clustering hierarchy and then determine the number of clusters by analyzing the lifetime of the hierarchy. The criteria used for this kind of evaluation are called "**internalcriteria**".

In agglomerative clustering algorithms, all segments of spectral wavelengths are equally suitable to be merged into a new single cluster, regardless of whether they are adjacent in the spectral wavelengths, the so-called "**global principle**". Thus, there is the possibility that two spectral bands with a certain interval in spectral wavelengths are merged into the same cluster. Combining the two principles, local and global, we can use some prior knowledge of spectral wavelengths to lower the dimensions of dissimilarity and reduce the computational complexity.

A diagram of the proposed method is shown in Fig. 4.9.

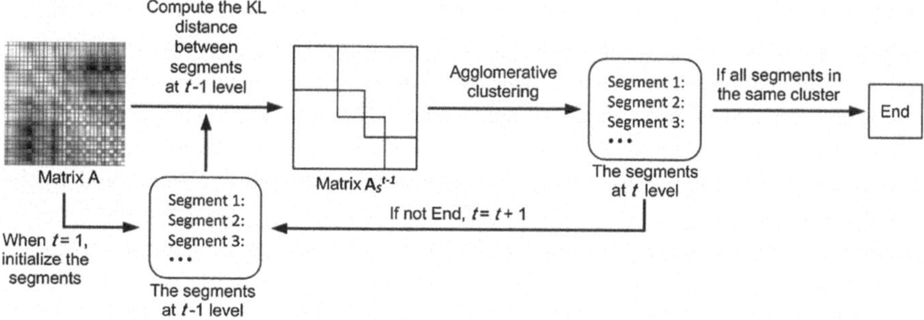

Fig. 4.8 A diagram of agglomerative clustering

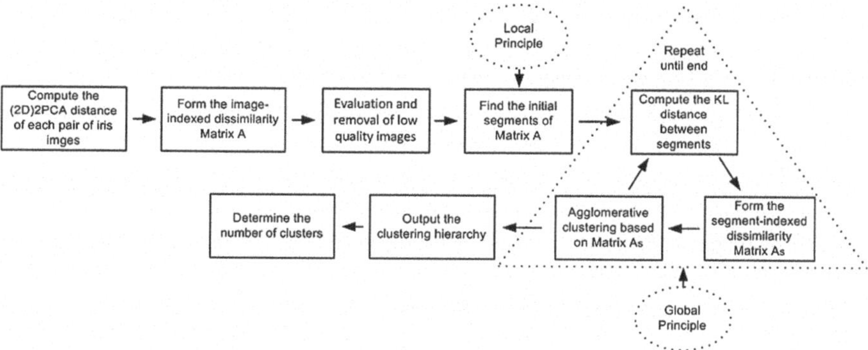

Fig. 4.9 A diagram of the proposed method

4.4 Experimental Results and Analysis

A dataset that contained samples from 60 irides was used to conduct the following study. Our iris database, which uses the proposed multispectral iris acquisition system, was created with 30 subjects. The volunteers are the students and staff at the Harbin Institute of Technology, Shenzhen Graduate School (HITSGS).

In this database, 25 are male, and the age distribution is as follows: younger than 30 years old comprise 80 %, and between 30 and 40 years old comprise about 20 %. Twenty images, which were taken from each the left and right eye of a subject, are selected under 12 spectral bands that correspond to 420, 490, 545, 590, 635, 665, 700, 730, 780, 810, 850, and 940 nm, respectively. In total, we collected 14,400 iris images for this database. The image resolution is 640 × 480.

Clustering the spectral wavelengths should be based on the maximum dissimilarity of iris textures captured in the corresponding spectral wavelengths, so only the iris rather than any other parts of the image (such as the pupil, sclera, and eyelid) can serve as the basis for clustering. In this work, we use normalized iris images to eliminate interference on the accuracy of clustering.

Daugman's rubber sheet model is clearly the most common in iris normalization methods and can be used to unwrap the iris from Cartesian coordinates to a pseudo-polar coordinate system. The mask that denotes the segmented iris is also unwrapped into this new coordinate system (see Fig. 4.10).

The ways in which iris texture may be corrupted, aside from motion blurs, camera noise, and inaccurate focusing, are occlusion by the eyelashes and eyelid (usually from the upper eyelashes and lower eyelid). These often have random and complex shapes, combining with each other to form masses of intersecting elements rather than just simple hairlike strands that might be amenable to detection by elementary shape models (Daugman 2007). As shown in Fig. 4.10, these detected occlusions in the lower eyelid (on the left side) and upper eyelashes (on the right side) within the iris have now been marked as white pixels. They can be the strongest signals on the iris image in terms of contrast or energy and dominate the texture content with spurious information. Even when they are detected and excluded from the iris texture, irreversible effects will be observed: The real iris texture is never to be recovered.

In this work, we take three measures to eliminate the interference of occlusion by eyelashes and the eyelid. First, we use only the lower half (180°) of the iris region to avoid being occluded by the upper eyelashes. Second, we ask the subjects to widely open their eyes to avoid occlusion by the lower eyelid. Finally, we have also hand-selected iris images in order to remove the images with occlusion from the eyelashes and eyelid. After normalization, only 10 spectral bands longer than 500 nm are used in the experiment; since in the 420 and 490 nm images, the gray scales of the pupil and iris are too close to be correctly segmented and the iris textural clarity is too low to benefit this study.

The distance based on $(2D)^2$PCA is relatively sensitive to pupil dilation, so we used 6 images with the smallest pupil of each spectral band, and in total, 3600 iris images are used for this experiment. The radius of the pupils in these images is very

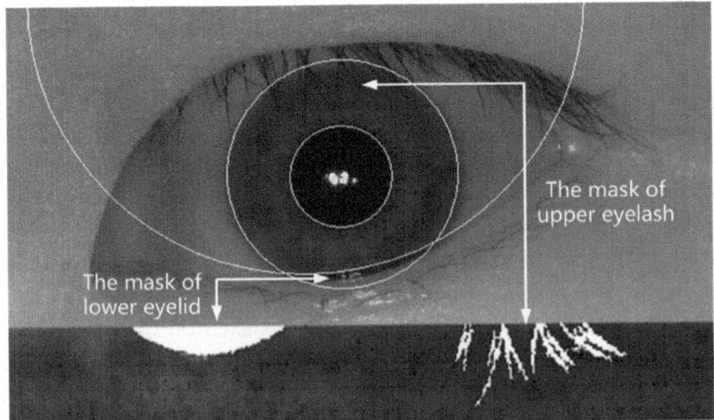

Fig. 4.10 The unwrapped iris image and the mask

small and changes only in a limited range of size. In consideration of the degree of iris deformation, we use the ratio of pupil radius to iris radius as $T = R_p/R_i$ to measure the influence of pupillary constriction. In this work, the T of all 3600 images is in the range of 0.23–0.35.

For each iris, the image-indexed dissimilarity Matrix **A** (60 × 60) is shown in Fig. 4.11. Regions of the highest similarity, generated from similar iris texture within the same spectral wavelength, appear as the brightest squares on the diagonal. The relatively brighter rectangular regions off the main diagonal indicate the similarity between the different spectral wavelengths. The region with near-infrared wavelengths that are longer than 800 nm is more self-similar than the other regions, which is shown in Fig. 4.11. This observation result is not surprising, if we take into account the difference of reflectance across all spectral wavelengths. The near-infrared illumination (especially that longer than 800 nm) produces almost no reflections. On the other hand, visible illumination produces reflectance more easily, which becomes the interference factor to similar texture in iris images and results in more dissimilarity.

We can obtain a series of novelty scores by correlating Matrix **A** with a checkerboard kernel and find the cutoff point of the spectral wavelength. A 6 × 6 checkerboard kernel with a radial Gaussian taper is used at the junction of the two adjacent spectral wavelengths rather than other places, based on the following assumption: Two images of the same spectral wavelength should not be divided into two different clusters. By comparing the novelty scores with a preset threshold, we get 6 cutoff points that divide the iris image sequence into 7 segments as shown in Table 4.1. Based on the cutoff points for segmentation, we can calculate a

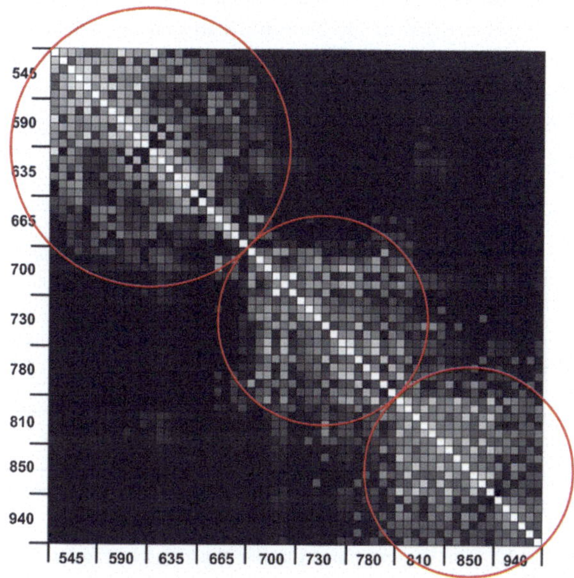

Fig. 4.11 The image-indexed dissimilarity Matrix **A**

Table 4.1 The 7 initial segments of the iris image sequence

Segment 1 (nm)	Segment 2 (nm)	Segment 3 (nm)	Segment 4 (nm)	Segment 5 (nm)	Segment 6 (nm)	Segment 7 (nm)
545	635	665	700	730	810	940
590				780	850	

dissimilarity matrix of substantially lower dimension, indexed by the number of segments instead of the number of images.

In accordance with Table 4.1, we consider these 7 initial segments as the input of the clustering algorithm, compute the inter-segment dissimilarity (the KL distance) between each pair of segments, embed them into the first segment-indexed dissimilarity Matrix A_s^0 (7 × 7), and use an agglomerative clustering algorithm to merge clusters for the first time. Then, we repeat the above process and get the dendrogram from the clustering hierarchy (see Fig. 4.12).

As explained earlier, this algorithm determines a whole hierarchy of spectral wavelength clustering, rather than a single clustering. However, in this work, we are only interested in the specific clustering that best fits the data. Thus, we have to decide which clustering of the produced hierarchy is most suitable for the data. Equivalently, we must determine the appropriate level to cut the dendrogram that corresponds to the clustering hierarchy.

First, the best clustering within a given hierarchy generated by an agglomerative algorithm should be determined. Clearly, this is equivalent to the identification of the number of clusters that best fits the data. An intuitive approach is to search in the proximity dendrogram for clusters that has a large lifetime, which is used as the "**internal criteria**" for cluster validity of the proposed method. The lifetime of a cluster is defined as the absolute value of the difference between the proximity level at which it is created and the proximity level at which it is absorbed into a large cluster (Theodoridis and Koutroumbas 2006; Everitt et al. 2001). According to the lifetime principle, the dendrogram in Fig. 4.12 suggests that three major clusters are enough for iris multispectral fusion.

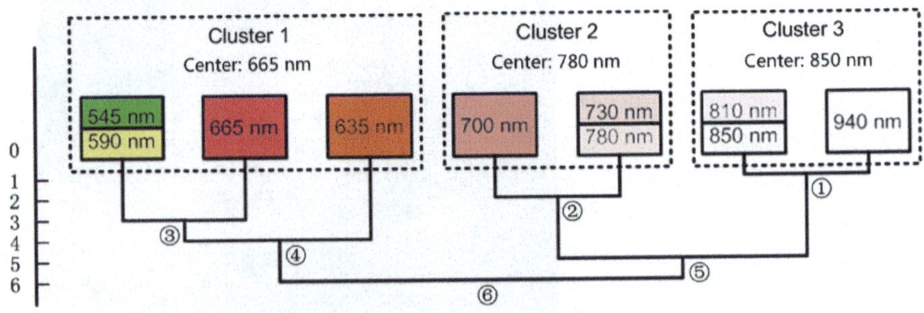

Fig. 4.12 The dendrogram from the clustering hierarchy

Second, we should find the cluster centers. We cannot directly derive three centers from the clustering hierarchy, because Clusters 2 and 3 are composed of only two nodes, respectively, from which the center of cluster cannot be distinguished. So, we will use the dissimilarity data in Matrix **A** to calculate the center of the corresponding cluster.

Suppose we have ***B*** bands of spectral wavelengths and ***S*** iris images from same subject captured under each band of spectral wavelengths. We denote by X_s^b, the sth iris image under the bth band. The distance $d_{\text{cross}}(b, b')$ between bands bth and b'th is defined as follows:

$$d_{\text{cross}}(b, b') = \frac{1}{S^2} \sum_{\substack{p=1 \\ q=1}}^{S} d^{b+b'}\left(X_p^b, X_q^{b'}\right) \tag{4.12}$$

and $d^{b+b'}$ is defined in Sect. 4.3, whose value is the elements of Matrix **A**. For a given band bth, we can calculate the average distance between bth and other bands in the same cluster as follows:

$$d_{\text{ave}}(b) = \frac{1}{N-1} \sum_{n=1}^{N-1} d_{\text{cross}}(b, n) \tag{4.13}$$

N is the number of spectral bands within this cluster. So, the center of cluster is as follows:

$$C = \{i | \arg \min d_{\text{ave}}(i)\}, \quad i = 1, 2, \ldots, N \tag{4.14}$$

Using this method, we can determine the three corresponding cluster centers. Note that in each run of the proposed method, the clustering results are based on the dataset that contains 60 images captured from one specific iris, so we should repeat the proposed method on the image dataset from different irides. The most frequent cluster number and centers are kept as the final result, because there is the possibility that the clustering results are influenced by the images of a special iris.

After testing all 3600 images from 60 irides, the final clustering hierarchy is the same as shown Fig. 4.12, and the three corresponding cluster centers are as follows: Cluster 1—665 nm, Cluster 2—780 nm, and Cluster 3—850 nm.

The validation of clustering accuracy can be obtained by exhaustively searching for the best partition of the spectral bands. In fact, under the constraint that clusters are composed of consecutive bands, there are only 15 distinct ways to partition the seven segments into three clusters, as shown in Table 4.2. So, the metric of clustering quality could be computed for each of the 15 combinations, and the best one could be used as a reference to compare with the clustering result of the proposed method.

Table 4.2 All 15 ways to partition the seven segments into three clusters

Combination	Cluster 1	Cluster 2	Cluster 3
1	1	2	3,4,5,6,7
2	1	2,3	4,5,6,7
3	1	2,3,4	5,6,7
4	1	2,3,4,5	6,7
5	1	2,3,4,5,6	7
6	1,2	3	4,5,6,7
7	1,2	3,4	5,6,7
8	1,2	3,4,5	6,7
9	1,2	3,4,5,6	7
10	1,2,3	4	5,6,7
11	1,2,3	4,5	6,7
12	1,2,3	4,5,6	7
13	1,2,3,4	5	6,7
14	1,2,3,4	5,6	7
15	1,2,3,4,5	6	7

For each of the 15 combinations, the dissimilarity (the KL distance) between each pair of clusters (C_i and C_j) can be computed as the following:

$$d_{\text{clu}}(i,j) = d_{\text{seg}}(p_i, p_j) = \exp\left(-\hat{d}_{\text{KL}}\left(N(\mu_i, \Sigma_i)\middle\|N(\mu_j, \Sigma_j)\right)\right) \tag{4.15}$$

Also, the dispersion of the cluster C_i is defined as the following:

$$S_i = \frac{1}{N}\sum_{j=1}^{N} d_{\text{ave}}(j) \tag{4.16}$$

N is the number of spectral bands within the cluster C_i, and $d_{\text{ave}}(j)$ is the average distance between j and other bands in the cluster C_i.

So the dissimilarity and the dispersion can be combined to output one clustering quality score Q for each combination as the following:

$$Q = \frac{S_1 + S_2}{d_{\text{clu}}(1,2)} + \frac{S_2 + S_3}{d_{\text{clu}}(2,3)} + \frac{S_1 + S_3}{d_{\text{clu}}(1,3)} \tag{4.17}$$

In this way, small values of Q are indicative of the presence of compact and well-separated clusters. The values of Q corresponding to 15 combinations are shown in Fig. 4.13, and we seek the minimum value of Q from them. In Fig. 4.13, the Combination 11 has the minimum value of Q. We should repeat the above process on the image dataset from different irides, and the most frequent combination corresponding to the minimum value of Q should be kept as the final result.

Fig. 4.13 Values of Q corresponding to 15 combinations

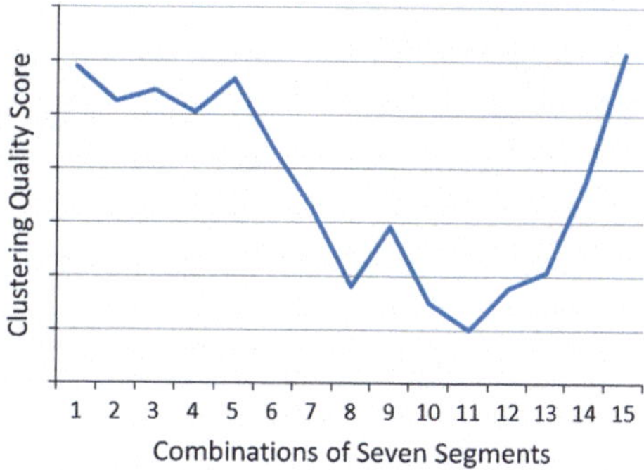

After testing all 3600 images from 60 irides, the Combination 11 had the highest frequency of occurrence (86.7 %), which verified the results of the proposed method: Cluster 1—from 545 to 665 nm, Cluster 2—from 700 to 780 nm, and Cluster 3—from 810 to 940 nm.

Now, we will analyze the interference factors and their impacts on the clustering. There are two interference factors that significantly influence the clustering results: The first is the changing radius of the pupil caused by pupil dilation and contraction, and the second is the reflectance of different portions of the spectral wavelengths.

First, we analyze the interference factor—the changing radius of the pupil. The pupil radius of each image can be calculated in iris segmentation, which is used as the selection criteria for experimental dataset, because the radius of the iris has been normalized to 100 pixels by the bilinear interpolation method. For one iris, the distribution range of the ten smallest pupil radiuses is a little bit different across all spectral wavelengths and a gradual range increase is observed with an increase in wavelengths, sometimes more than 10 pixels (see Fig. 4.14). This proved even with the flickering checkerboard that causes pupil dilation and contraction, and the intensity of multispectral illumination still has different degrees of stimulation on the pupil. Generally, the illumination from 545 to 700 nm is obviously visible, so

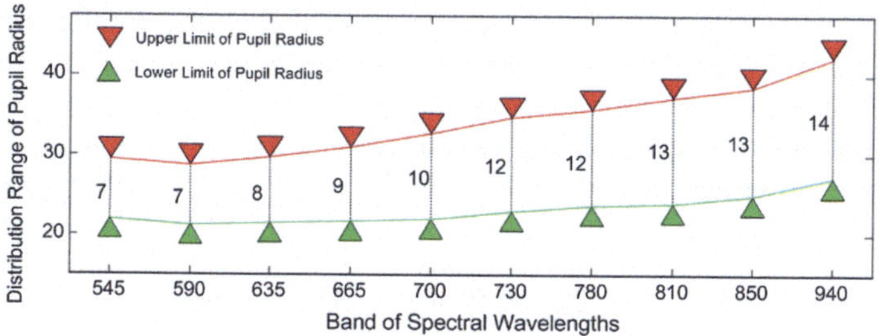

Fig. 4.14 Analysis of the first interference factor—changing pupil radius: distribution range of the ten smallest pupil radiuses across all spectral wavelengths. The shorter wavelength spectral band has a relatively smaller pupil radius than the one of longer spectral wavelengths

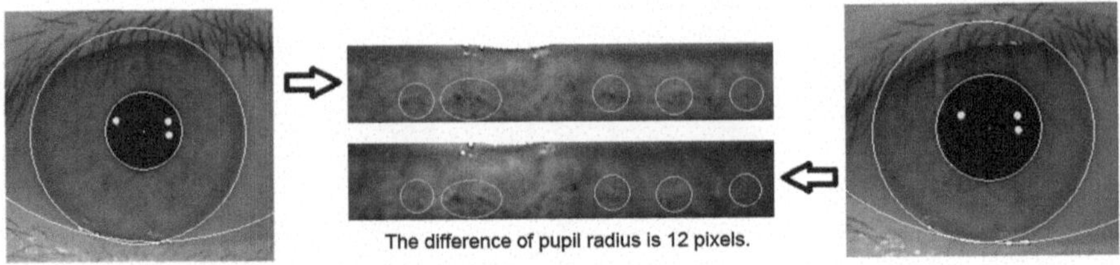

Fig. 4.15 Analysis of the first interference factor—changing pupil radius: a comparison of two normalized images with different radiuses of pupil. The different unwrapped texture caused by changing pupil radius is marked with *circles*

the distribution range of the pupil radius is relatively small and concentrated, while the illumination from 700 to 940 nm is weakly visible, so the distribution range of the pupil radius is relatively large and scattered. The distribution range of the pupil radius can be considered as the measurement of nonlinear texture deformation, which cannot be effectively corrected by Daugman's rubber sheet model and will become the interference to the accuracy of clustering. Figure 4.15 is a comparison of two iris normalized images, whose difference in pupil radius is 12 pixels, and we can observe the differences in texture marked with circles, which is not very obvious but can still be seen. Hence, the shorter wavelength spectral band introduces less nonlinear texture deformation and is more likely to form a compact cluster than a band of longer spectral wavelengths, at least in theory.

The average distance between image samples within each band of spectral wavelengths can be considered as the indirect measure of the compact cluster, which varies from each other and presented by two curves that correspond to the two methods (see Fig. 4.16). The general trend is clear: Shorter wavelength spectral bands have a relatively larger average distance than the longer wavelength spectral bands.

These experimental results seem inconsistent with the theoretical assumption made from the previous texture deformation analysis, which can be explained by two reasons. One reason is that different texture is generated by the structure or melanin across all spectral wavelengths, and the other reason is attributed to the second interference, that is, spectral reflectance.

From an observation of iris images, we found that the shorter wavelength spectral band will introduce more obvious reflectance, even extracted as the most texture information. Due to the reflected images of different surrounding objects on the iris surface, the shape and location of reflectance has a certain randomness, which will result in unexpected dissimilarity between the iris images. In Fig. 4.17, two iris images captured in the same band of 635 nm have different reflectance, which are marked with arrows and obvious as to show some discrimination. Other reflectance is located in the upper half of the iris, outside the region of the normalization, so they cannot introduce interference.

Due to the reflection properties of the eye surface, in the band of longer spectral wavelengths, such as NIR, less reflectance will be captured in iris images than in the

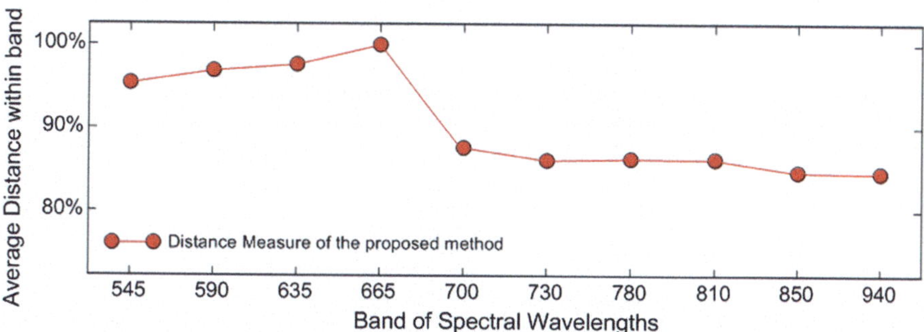

Fig. 4.16 Analysis of the second interference factor—spectral reflectance: a comparison of average distance within bands across all spectral wavelengths. Shorter wavelength spectral bands have a relatively larger average distance than the longer spectral wavelength bands

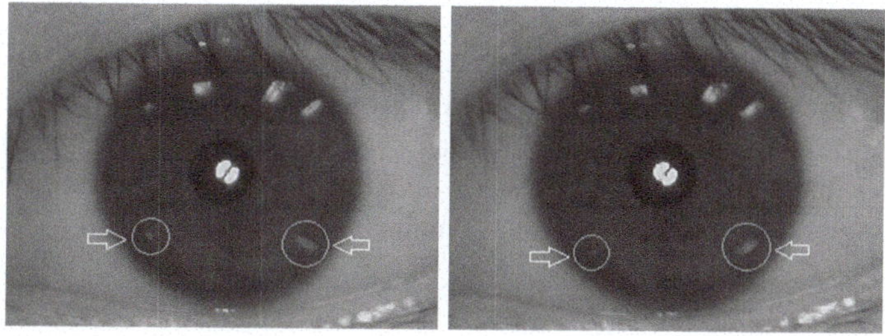

Fig. 4.17 Analysis of the second interference factor—spectral reflectance: a comparison of 635 nm iris images with different reflectance marked with *arrows*

band of shorter ones. This means that without the interference of reflectance, the longer wavelength spectral band is more likely to form a compact cluster than the shorter wavelength spectral band, which is consistent with previous experimental results. Hence, the low-quality images caused by reflectance of short spectral wavelengths may become the main threat to the accuracy of iris multispectral fusion.

Thus, taking into account the joint effect of interference factors, we know that the clustering result of spectral wavelengths is the result of complex interactions of many influences—including the changing pupil radius, varying spectral reflectance, and iris textures generated from the structure or melanin (see Fig. 4.18). The above analysis is important to the practical application of iris multispectral fusion, because all the factors previously mentioned actually exist and their impacts cannot be ignored. In fact, band selection should be done with caution, especially from Cluster 1, since the introduction of serious random reflectance will greatly influence the matching performance.

Finally, we can find some physical explanation for the clustering result. The spectral wavelengths longer than 800 nm can be seen as the typical NIR spectrum, and ones that are shorter than 700 nm are considered as visible light. The wavelengths

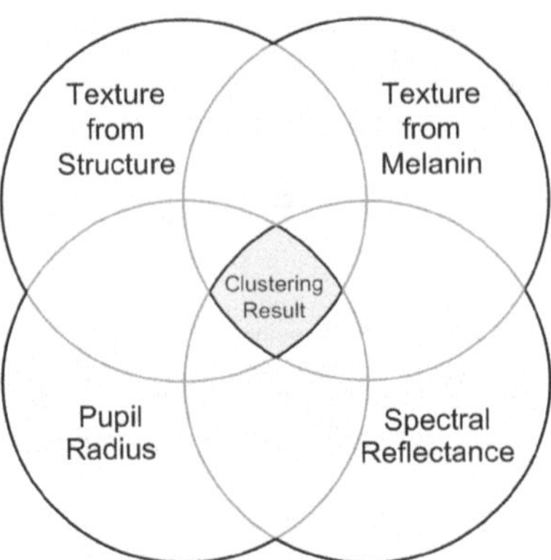

Fig. 4.18 The four factors that influence the clustering of spectral wavelengths: two main factors—texture from structure and melanin, and two interference factors—pupil radius and spectral reflectance

from 700 to 800 nm are the junction area between visible light and NIR, near the upper limit for response by a typical human eye, which contain both a certain proportion of visible light and NIR. The above division of wavelengths is consistent with the clustering result, which proves our conclusion has some physical basis.

4.5 Summary

This work uses East Asian irides as research subjects and explores the possibility of clustering spectral wavelengths from 420 to 940 nm based on the maximum dissimilarity of iris textures captured in the corresponding spectral wavelengths, which will provide an important standard for selecting spectral wavelength bands for iris multispectral fusion. The eventual goal is to present the number of spectral wavelength bands that will be enough for black-based iris multispectral fusion and determine these bands. This represents the first attempt in the literature to analyze the black irides of East Asians in a multispectrum analysis.

Such an analysis is required to understand the texture from the structure and melanin of the iris that is revealed across all wavelengths from 400 to 1000 nm. So, an acquisition system that is combined with a checkerboard stimulus has been designed to acquire a small dataset of East Asian black iris images in the 420–940 nm wavelength range. The checkerboard stimulus is specially designed for the acquisition of iris images with pupillary constriction, and a similar stimulus is used for the first time in a multispectral iris acquisition system.

In this work, we use agglomerative clustering based on multigroup $(2D)^2$PCA, which is improved to analyze group elements, such as the images under different bands of spectral wavelengths.

The initial experimental results suggest that 3 clusters are enough to represent the 10 feature bands of spectral wavelengths from 545 to 940 nm: Cluster 1 is from 545 to 665 nm, Cluster 2 is from 700 to 780 cm, and Cluster 3 is from 810 to 950 nm. The centers of Clusters 1, 2, and 3 are 665, 780, and 850 nm, respectively. These results are also verified by the k-means clustering method based on Hamming distance.

A further analysis shows that the clustering results are a complex mixture of two joint effects: One is the anatomical difference in the iris structure and melanin revealed in these wavelengths; the other is the interference factor, such as varying pupil radius and spectral reflectance. Finally, we have evaluated the potential impact of interference on the performance of iris multispectral fusion.

In the future, fusion at the image and feature levels should be investigated, based on the iris images that are simultaneously acquired, which has attracted the attention of some researchers (Vilaseca et al. 2006, 2008). More effective enhancement techniques for image quality may be explored to improve the performance of multispectral iris recognition in large datasets (Vatsa et al. 2008).

References

Baraniuk RG, Jones DL (1993) A signal-dependent time-frequency representation: optimal kernel design. IEEE Trans Signal Process 41(4):1589–1602

Boyce CK (2006) Multispectral iris recognition analysis: techniques and evaluation. Doctoral dissertation, West Virginia University

Burge MJ, Monaco MK (2009) Multispectral iris fusion for enhancement, interoperability, and cross wavelength matching. In: SPIE defense, security, and sensing, pp 73341D

Cover TM, Thomas JA (1991) Elements of information theory. Wiley, New York

Daugman JG (1993) High confidence visual recognition of persons by a test of statistical independence. IEEE Trans Pattern Anal Mach Intell 15(11):1148–1161

Daugman J (2007) New methods in iris recognition. IEEE Trans Syst Man Cybern B 37(5):1167–1175

Everitt B, Landau S, Leese M (2001) Cluster analysis, 4th edn. Edward Arnold, London

Foote J (2000) Automatic audio segmentation using a measure of audio novelty. In: IEEE international conference on multimedia and expo, vol 1. IEEE, Piscataway. pp 452–455

Franssen L, Coppens JE, van den Berg TJ (2008) Grading of iris color with an extended photographic reference set. J Optom 1(1):36–40

Ghosal S, Van Der Vaart AW (2001) Entropies and rates of convergence for maximum likelihood and Bayes estimation for mixtures of normal densities. Ann Stat 1233–1263

Ghosal S, Van Der Vaart A (2007) Posterior convergence rates of Dirichlet mixtures at smooth densities. Ann Stat 35(2):697–723

Ghosal S, Ghosh JK, Ramamoorthi RV (1999) Posterior consistency of Dirichlet mixtures in density estimation. Ann Stat 27(1):143–158

Hollingsworth K, Bowyer KW, Flynn PJ (2009) Pupil dilation degrades iris biometric performance. Comput Vis Image Underst 113(1):150–157

Johnson DH, Sinanovi S (2001) Symmetrizing the Kullback-Leibler distance. IEEE Trans Inf Theory

Li Z, Sun F (2005) Pupillary response induced by stereoscopic stimuli. Exp Brain Res 160(3):394–397

Liu Z, Yan JQ, Zhang D, Li QL (2007) Automated tongue segmentation in hyperspectral images for medicine. Appl Opt 46(34):8328–8334

Ozawa K (1983) CLASSIC: a hierarchical clustering algorithm based on asymmetric similarities. Pattern Recognit 16(2):201–211

Park JH, Kang MG (2007) Multispectral iris authentication system against counterfeit attack using gradient-based image fusion. Opt Eng 46(11):117003

Ross A, Pasula R, Hornak L (2006) Exploring multispectral iris recognition beyond 900 nm. In: Proceedings of the 2006 conference on computer vision and pattern recognition workshop, p 51

Sun FC, Chen LY, Zhao XZ (1998) Pupillary responses evoked by spatial patterns. Acta Physiol Sin 50(1):67–74

Theodoridis S, Koutroumbas K (2006) Pattern recognition, 3rd edn. Elsevier, Amsterdam

Thornton J, Savvides M, Kumar V (2007) A Bayesian approach to deformed pattern matching of iris images. IEEE Trans Pattern Anal Mach Intell 29(4):596–606

Tokdar ST (2006) Posterior consistency of Dirichlet location-scale mixture of normals in density estimation and regression. Sankhya Indian J Stat 67: 90–110

Vatsa M, Singh R, Noore A (2008) Improving iris recognition performance using segmentation, quality enhancement, match score fusion, and indexing. IEEE Trans Syst Man Cybern B Cybern 38(4):1021–1035

Vilaseca M, Pujol J, Arjona M, de Lasarte M (2006) Multispectral system for reflectance reconstruction in the near-infrared region. Appl Opt 45(18):4241–4253

Vilaseca M, Mercadal R, Pujol J, Arjona M, de Lasarte M, Huertas R, Melgosa M, Imai FH (2008) Characterization of the human iris spectral reflectance with a multispectral imaging system. Appl Opt 47(30):5622–5630

Wilkerson CL, Syed NA, Fisher MR, Robinson NL, Albert DM (1996) Melanocytes and iris color: light microscopic findings. Arch Ophthalmol 114(4):437–442

Zuo W, Zhang D, Wang K (2006) Bidirectional PCA with assembled matrix distance metric for image recognition. IEEE Trans Syst Man Cybern B Cybern 36(4):863–872

Chapter 5
The Prototype Design of Multispectral Iris Recognition System

Abstract Compared with two traditional designs "two-camera" and "one-camera", our proposed design method has some obvious advantages. The proposed dual-eye capture device benefits from simplicity and lower cost, because of reduced optics, sensors, and computational needs. This system is designed to have good performance at a reasonable price so that it becomes suitable for civilian personal identification applications. This research represents the first attempt in the literature to design the dual-eye multispectral iris capture handheld system based on single low-resolution camera.

Keywords Mutlispectral iris · Dual-eye capture · 1-D log-Gabor · Hamming distance

5.1 Introduction

A critical step in an iris recognition system is designing an iris capture device that can capture iris images in a short time. Some research groups (Wildes 1997; Park and Kim 2005; Tan et al. 1999; CASIA Iris Image Database 2005; Shi et al. 2003), such as OKI, LG, Panasonic, and Cross-match, have explored the requirements on the iris image acquisition system, and some implementations have already been put into commercial practice (Biom Technol 2005; He et al. 2008; Wilkerson et al. 1996). Some research groups (Ross et al. 2006; Vilaseca et al. 2008; Burge and Monaco 2009; Ngo et al. 2009; Gong et al. 2012a, b) explored the single-eye multispectral iris image capture devices, which are designed for the experimental data collection only, far from the requirements of real usage scenarios. All of the previous multispectral devices are designed for capturing iris image from single eye each time. It switches the light source or the filter manually, adjust the lens focal length manually, and use chinrest to require an uncomfortable fixed head position. To some extent, most of multispectral devices demand full cooperation from the subject who needs to be trained in advance, which will eventually increase the time

D. Zhang et al., *Multispectral Biometrics*,
DOI 10.1007/978-3-319-22485-5_5

of image acquisition and influence the acceptability of users. Subjects may be tired and easily fatigued, blinking eyes, rotating eyes, dilating, and constricting pupil subconsciously, which are interference factors to iris image quality. Taking into account these interference factors in iris image acquisition, the acquisition of only an iris cannot ensure the quality of the iris image, and the accuracy of the iris recognition is destroyed, especially in some application scenarios of higher security requirements, combining left and right irises recognition for personal authentication. So, the practical system should be designed to capture iris images from two eyes jointly, to guarantee at least the iris image of one side is of high quality.

Although single-eye systems benefit from simplicity and lower cost thanks to reduced optics, sensors, and computational needs, but individual scanning implies a sequential activity that increases enrollment time and risks left–right misassignment. Dual-eye systems are more complex because a larger space must be imaged, processed, and segmented to isolate each iris, but there are also very obvious advantages: Enrollment throughput can be faster, with less risk of left–right iris swapping, more flexibility in obtaining individualized, best-quality iris images from any one of the two eyes.

Using dual-eye system, two iris codes from left and right eyes are not fused: The code from left eye is compared exhaustively against all in left eye database, and the code from right eye is compared exhaustively against all in right eye database. So, two recognition decisions d_{left} and d_{right} are made from two eyes, respectively, and the following two final decision-making policies depend on the level of security users choose: Policy 1—the identity of individual is determined by any one of two eyes, $d_{\mathrm{final}} = d_{\mathrm{left}}$ OR d_{right}; Policy 2—the identity of individual is determined by both two eyes, $d_{\mathrm{final}} = d_{\mathrm{left}}$ AND d_{right}. The probability of misidentification when using Policy 2 will be reduced by many orders of magnitude.

Some commercial groups (Mobile Dual Iris Capture Device 2014; IriMagic™ Series 2000; Dual Iris Scanner 2014) explored the handheld dual-eye iris image capture devices as in Table 5.1. All of the previous dual-eye devices are based on two cameras and designed for capturing iris image under only one band of wavelength, without the feature of multispectral image capture, so it cannot be used to multispectral iris recognition.

Simple dual-eye capture device can be done with two single-eye cameras connected together using a bracket of some kind, see Fig. 5.1. Two slide bars mounted on a tripod keep the lens axes parallel, and avoid rotation as you adjust the cameras between the first and second iris images. For best image quality, shutter controls should be added so both cameras could be triggered at the same time. Because of slight variations among cameras shutter speeds and exposures, we cannot guarantee that the image quality (such as brightness, contrast, gray level distribution) of the two iris images is exactly the same. The downside is that you have to carry the slide bar and a tripod with you, and the process of system calibration (including the magnification and focus accuracy of two sets of lens, parallel degree of main optical axis) is so cumbersome as to reduce the efficiency of image acquisition. The "two-camera" structure is too bulky and cannot be directly used for handheld iris capture devices.

Table 5.1 The comparison of previous handheld dual-eye capture devices

	Manufacturer	Mode	Number of IRIS	Wavelength band switching	Focusing mode	Capturing mode	Number of camera	Power
1	CrossMatch, USA	I SCAN 2	Dual-eye	No	Manually	Contact	Two	Single USB
2	Iritech, USA	BD300	Dual-eye	No	Manually	Non-contact	Two	Dual USB
3	Cogent, USA	CIS 202	Dual-eye	No	Manually	Contact	Two	Single USB
4	CML, Korea	DMX-10	Dual-eye	No	Manually	Non-contact	Two	5VDC2A adaptor
5	Hongbog, Korea	Ueye-D D1	Dual-eye	No	Manually	Non-contact	Two	Single USB

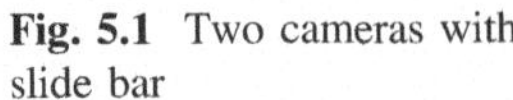

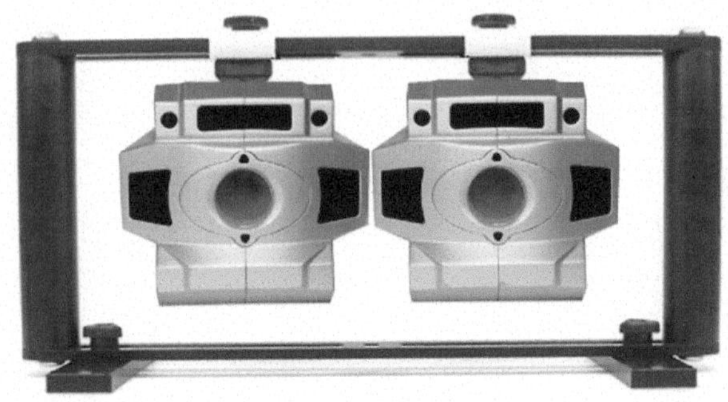

Fig. 5.1 Two cameras with slide bar

So, the previous dual-eye devices in Table 5.1 are all based on the optimized similar design, but in more compact structure.

More advanced dual-eye capture device can be done with only one camera. The iris recognition standards (ISO/IEC, 2004) recommend 200 pixels across the iris as the minimum acceptable resolution for level 2 iris features (intermediate iris structure sufficient to identify prominent instances of radial furrows, concentric furrows, pigment spots, crypts of Fuchs, and degenerations due to medical conditions or damage). At this writing, the highest resolution commercial video cameras are of the order of 4096 × 4096 pixels (16 million pixels), so they can support a field of view of approximately 20 × 20 cm, only just able to cover the width of the two eyes with a very small margin on left and right sides. The "one-camera" design is only used in large iris access control system in train stations, libraries, airports, such as Sarnoff Company's Iris on the Move® (IOM). The "one-camera" design is too expensive to be acceptable on a very large scale in the emerging handheld iris capture device markets.

In order to avoid the design problems noted above, we abandoned the "two-camera" design and improved the "one-camera" design. Using a set of refraction lens, we enlarged the effective resolution of 2 million pixels camera, which can cover the width of the two eyes within 1600 × 1200 pixels, capture two iris images with high resolution and high image quality, and meet the iris recognition standards ISO/IEC 2004 (200 pixels across the iris—the minimum acceptable resolution for level 2 iris features). We presented a dual-eye multispectral iris capture handheld system that will enable collection of two iris images simultaneously based on single camera with 2 million pixels, to explore the feasibility of dual-eye multispectral capture handheld system with high efficiency and low cost. Using this system, a complete capture cycle (including six images from left and right iris under three different wavelength bands) can be completed within 2 or 3 s, much faster than previous devices. In addition, this system is not only an isolated acquisition device, but also connects to the server running recognition algorithm, so it can complete a full process of multispectral online iris identification, which is the first attempt of practical application of the multispectral dual iris recognition.

The capture system consists of the following four parts: (1) capture unit; (2) illumination unit; (3) interaction unit; and (4) control unit. It uses a Micron

1/3.2-in. CMOS camera, an automatic focusing lens and a set of refraction lens as the capture unit, and the working distance is about 160–200 mm. Two groups of matrix-arrayed LEDs (Light Emitting Diode) across three different wavelengths (including visible light and near-infrared light) are used as the multispectral illumination unit. We design an interaction unit including the infrared distance measuring sensor and the speaker to realize the exact focusing range of the lens via the real-time feedback of the voice prompts.

Compared with the traditional "two-camera" design, the proposed design has three advantages: (1) No synchronization control between two cameras, so the circuit board design is simpler, smaller, and higher reliability; (2) Using one camera rather than two, so more energy efficient, longer battery life, and more easily converted into a handheld portable device. Some previous "two-camera" capture devices, such as Iritech's DB300 and CMI's MDX-10 in Table 5.1, whose power consumption is much higher than the upper limit of one USB port, even need to be powered by two USB ports or external DC power supply. Without the external fixed power, such device should be completely unusable. (3) No need for cumbersome two-camera calibration (including the magnification and focus accuracy of two set of lens, parallel degree of main optical axis, brightness, contrast, shutter speeds, and exposure levels), the imaging parameters corresponding to left and right eye are exactly the same.

Compared with the traditional "one-camera" design, the proposed design has three advantages: (1) Using the low-resolution camera, so the system cost is significantly lowered; (2) Under conditions of limited transmission bandwidth, low-resolution video streams have higher frame rates, so more conducive to capture iris images from moving eyes and more resistant to eye movement interference, such as blinking eyes, rotating eyes, dilating, and constricting pupil subconsciously; (3) Based on Daugman's method, the process of detecting and localizing the eye, the iris, and the eyelids has the largest time consumption, almost 40 % of the total iris processing time. Using the same iris detection and location algorithm and the same computing platform, the computation time consumed by the low-resolution iris images will be significantly less (the complexity of detection and location is directly related to the size of iris image), so the real-time performance of the iris recognition system will be greatly improved.

In summary, compared with two traditional designs "two-camera" and "one-camera" , our proposed design method has some obvious advantages. The proposed dual-eye capture device benefits from simplicity and lower cost, because of reduced optics, sensors, and computational needs. This system is designed to have good performance at a reasonable price so that it becomes suitable for civilian personal identification applications. This research represents the first attempt in the literature to design the dual-eye multispectral iris capture handheld system based on single low-resolution camera.

5.2 System Framework

5.2.1 Overall Design

One of the major challenges of multispectral dual iris capture system is capturing two iris images of the left and right iris while switching wavelength band of illumination. The realization of the capture device is quite complicated, for it integrates light, machine, and electronics, into one platform and involves multiple processes of design and manufacture.

A multispectral iris image capture system is proposed. The design of the capture device includes following four subcomponents: (1) capture unit; (2) illumination unit; (3) interaction unit; and (4) control unit. The capture, illumination, and interaction units constitute the main body of the capture device, which is installed on the 3-way pan-tilt head of a tripod to allow subjects to manually adjust the pitching angle for fitting their height while capturing. The control unit is operating within the capture system, in charge of the synchronization of other three units, and data exchange with the iris recognition server.

We considered the configurations for our system as flowing: use one single camera with multiple narrow-band illuminators. The illuminators are controlled by ARM (Advanced RISC Machines) mail-board and can switch automatically and synchronize with the lens' focus and CMOS camera's shutter. This approach enables dual-eye collection of multispectral iris images.

The optical path is as follows: First, the subject is watching the capture window, and the multispectral light from the matrix-arrayed illuminators on both sides is delivered to the eyes. The reflected light from the subject's eyes is collected through a set of refraction lens, through a narrow-band filter, and imaged by the Micron CMOS camera using AF lens. In addition, the light beam from IR distance sensor is also concentrated on the focal plane, where the subject's eyes are located. The optical path of the proposed capture device is shown in Fig. 5.2a, b. The capture unit that we propose is composed of 4 parts: Micron CMOS sensor, AF lens, refraction lens, narrow-band filter, and protection glass, as shown in Fig. 5.3.

The camera with the Micron MT9D131 CMOS sensor is using USB interfaces and has exceptional features, including high resolution (working at 1600×1200), high sensitivity (Responsivity = 1.0 V/lux-sec), and high frame rate (15 fps at full resolution), which are all important to multispectral imaging. The CMOS's spectral response from 400- to 1000-nm wavelengths, including visible light and infrared light, is not absolutely uniform as in Fig. 5.4, but we verified that the CMOS response does not introduce significant errors into the experimental values after the optimization of the multispectral system.

In iris capture system, the acquisition of iris images almost always begins in poor focus. It is therefore desirable to compute focus scores for image frames very rapidly, to control a moving lens element for auto-focusing. AF lens is based on New Scale M3-F Focus Module (VistaEY2H Handheld Dual Iris Camera 1998) and controlled by the computer via I^2C interface, which running the 2-D fast focus

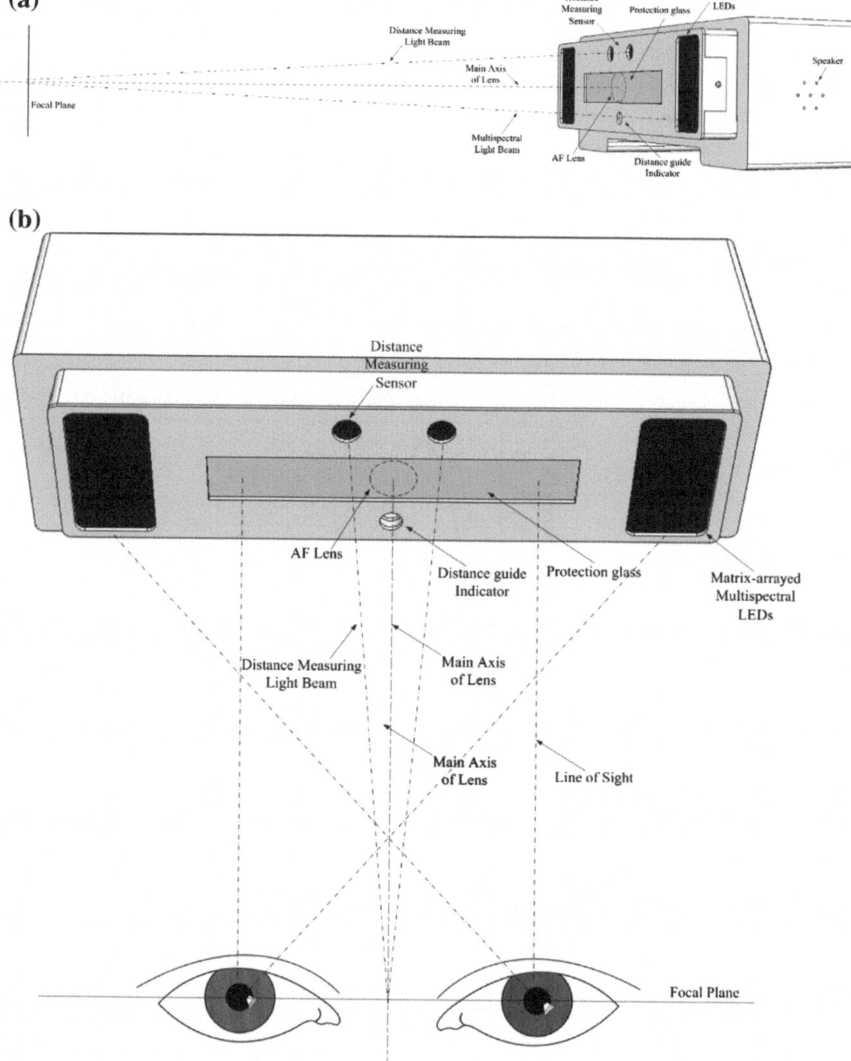

Fig. 5.2 The optical path. **a** Side view, **b** front view

assessment algorithm to estimate the quality of focus of a multispectral image and to indicate the direction of lens focus's movement. The focus range of the AF lens is set from 160 to 200 mm, so we can capture iris image that is in acceptably sharp focus in this distance range, without the strict requirements on the subject's location and cooperation.

The set of refraction lens is designed for capturing two iris images from left and right eye simultaneously and ensure that each iris image has enough effective

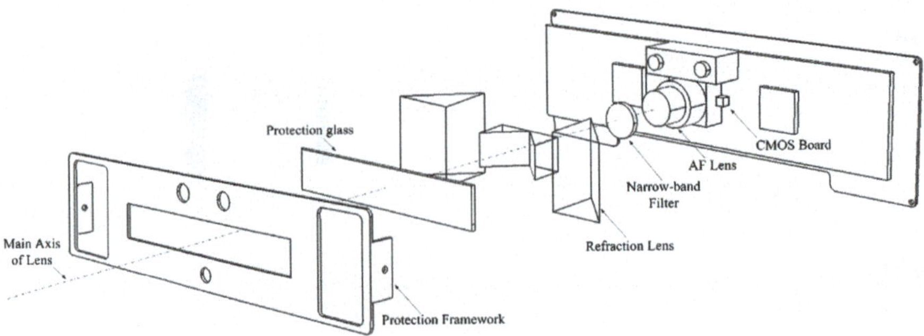

Fig. 5.3 The composition of capture unit

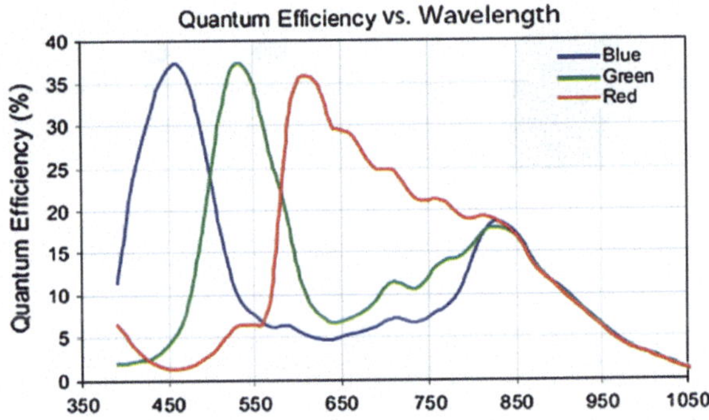

Fig. 5.4 Relative sensor response for the Micron MT9D131

resolution. The standards (ISO/IEC 19794-6: Information technology—Biometric data interchange formats—Iris image data) recommend a resolution of more than 200 pixels across the eye and demand at least 100 pixels. Many experiments have demonstrated that a reduction in the effective resolution of an iris image will increase the Hamming distance between the reduced resolution image and a full resolution image of the same eye (Matey et al. 2006). Effective resolution is affected by the resolution of camera and the focal length of Lens. Maximum resolution of Micron CMOS camera is 1600 × 1200. Following the standards of 200 pixels across the eye, one image can only cover the width of 80 mm. Generally, the inter-pupillary distance is 65–70 mm, and the iris diameter is 12 mm, so taking into account the necessary margin, the 80 mm is not enough to cover the area of two eyes.

We use a set of refraction lens to enlarge the effective resolution of iris area in the image as shown in Fig. 5.5. The refraction lens is made up of 4 triangular prisms (including 2 central prisms and 2 lateral prisms) and the prism housing. The central prisms must remain in contact with each other, so as to collect two optical paths (each one corresponding to an iris) and combine them into a light bundle of a few millimeters in diameter, and deliver it to the surface of CMOS via AF lens as in Fig. 5.5a. The lateral prisms, instead, must follow the average inter-pupillary

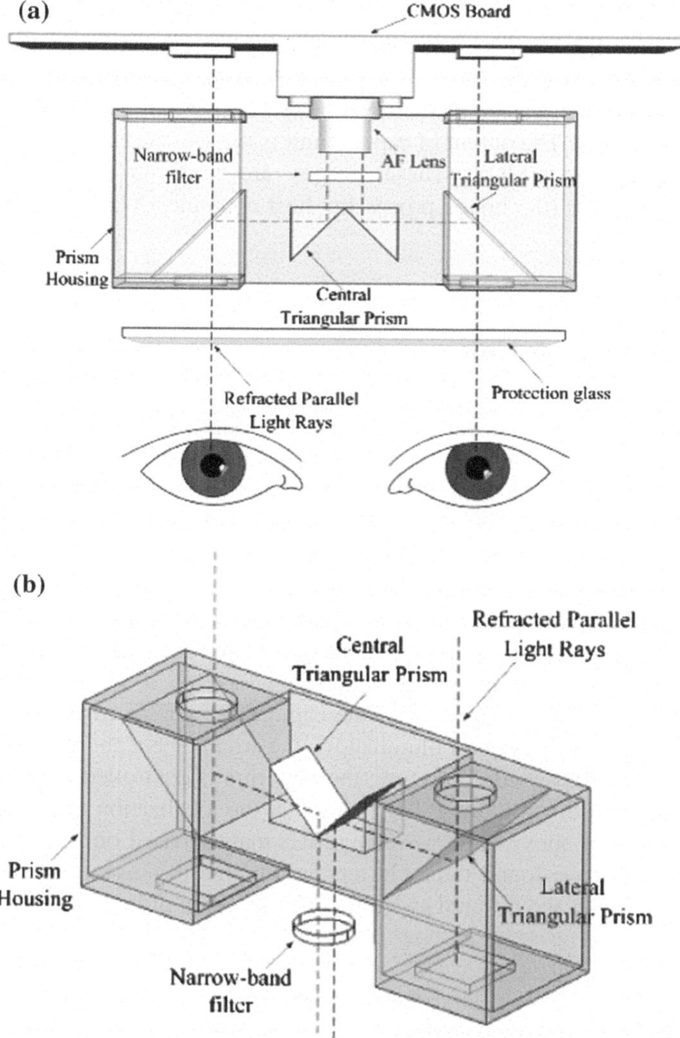

Fig. 5.5 Refraction lens. **a** Optical scheme of the refraction lens, **b** wiremesh view of the prism housing

distance, which is preset as 65 mm in this work, to aim at left and right eye accurately. The area between two eyes, such as nose, has not been acquired into the image to save the limited resolution. The image width of 1800 pixel is divided equally into two parts, and in each part the iris is located in the center. Each prism is mounted inside a short square tube which slides inside the main tube of the prism housing. Figure 5.5b illustrates the prism housing structure and its contents. The screws, support plates, and linkage tube have been omitted for clarity.

The narrow-band filter is the customized coated filter corresponding to the wavelengths of illuminators. The narrow-band filter can transmit the light of specified three wavelengths and reflect the light of other wavelengths. These three wavelengths with high transmittance are 700, 780, and 850 nm, under which the iris is imaged by CMOS. The output of capture unit is the sequence of multispectral iris images received by CMOS. These images are transmitted from the ARM main-board to server (the image processing host computer) via USB 2.0 interface for iris recognition.

5.2.2 *Illumination Unit*

The illumination system is composed of two groups, which are located at the bottom side of the lens, and each group includes nine matrix-arrayed (3 × 3) LEDs corresponding to three wavelengths: 700, 780, and 850 nm as shown in Fig. 5.6. The wavelengths of illuminators can be switched automatically, allowing illumination of the captured iris with a certain angle. The light of left LED is delivered to right iris, and the light of right LED delivered to left iris, which can increase the lighting angle of incidence, in order to reduce the interference of obvious reflectance in glasses or sunglasses. In the real usage environment, users often try to pass the iris recognition wearing glasses and sunglasses, so it is very necessary to improve anti-reflection interference of the capture device. To avoid glittering further, we optimized the design of illuminators: the arrangement and the angle of each LED are specially designed. Therefore, the glittering is controlled in the pupil area in most cases, and this glittering does not affect the localization and recognition.

The selection of above three wavelengths is mainly based on two reasons. The first reason is that we found three clusters are enough to present all wavelengths, including the visible and infrared spectrum. The selection of three wavelengths 700,

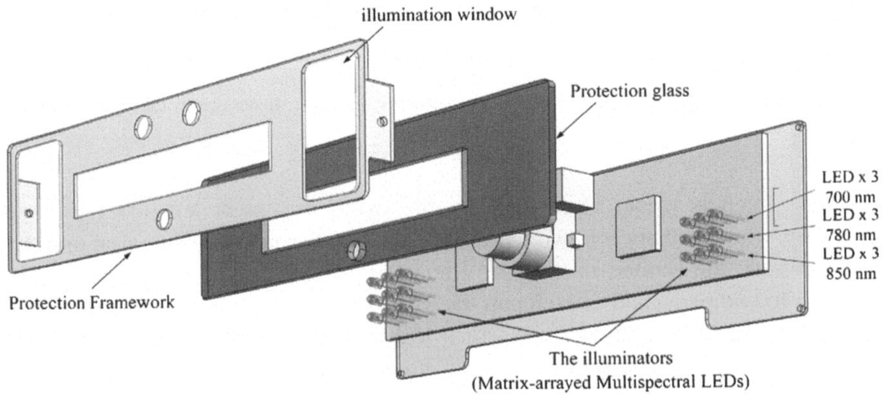

Fig. 5.6 The composition of illumination unit

780, and 850 nm is the optimized result, in order to ensure adequate coverage of the full spectrum and diversity of the iris texture. The second reason is that the practical considerations. The proposed iris capture device in this work will be developed into the practical multispectral dual iris recognition system, not just the experimental device, so all the selected wavelengths should be easily accessible in system. Selection of more wavelengths maybe means a little more accurate performance of iris recognition, but also results in much longer time of image acquisition and much lower acceptability of users. In summary, the selection of above three wavelengths is the best choice based on both experimental data integrity and feasibility of the practical system. In addition, the luminance levels for the experimental setup described here meet the requirements found in the ANSI/IESNA RP-27.1-05 for exposure limits under the condition of weak aversion stimulus.

5.2.3 Interaction Unit

The interaction unit is included in our iris capture device system for the purpose of easily capturing iris images. It is very convenient for subjects to adjust their poses according to the feedback information. The interaction unit that we propose is composed of 4 parts: checkerboard stimulus OLED, infrared distance measuring sensor, distance guide indicator, and the speaker, as in Fig. 5.7.

Pupil dilation is the most common quality defects in iris images, and serious interference factor especially for multispectral iris fusion and recognition. Hollingsworth et al. (2009) studied the effect of texture deformations caused by pupil dilation on the accuracy of iris biometrics and found that when matching two iris images (enrollment and recognition) of the same person, larger differences in pupil dilation yield higher template dissimilarities, and therefore, a greater chance of a false non-match. The above research demonstrated that texture nonlinear deformations caused by pupil dilation are sometimes so serious as to let iris recognition algorithms make this wrong decision.

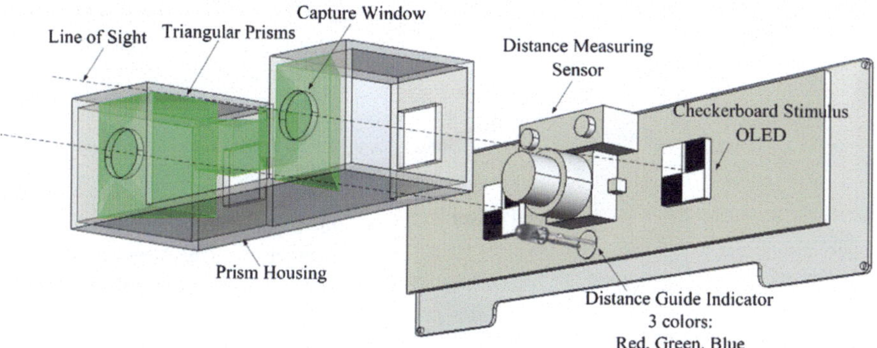

Fig. 5.7 The composition of interaction unit

In the proposed device, we design the checkerboard stimulus to solve the problem of pupil dilation. Some researcher (Sun et al. 1998) found that human pupillary constriction can be evoked by visual spatial patterns, such as the checkerboard. In the proposed device, the checkerboard is an OLED screen, on which the reversal pattern is a 2 × 2 matrix, can flicker between black on white to white on black (invert contrast) at a certain frequency without changing the illumination intensity on the eyes. When subject is watching the checkerboard stimulus, the pupillary constriction will be evoked. We capture the multispectral iris images with a similar degree of pupil dilation to minimize the interference caused by pupil dilation. The iris images of pupillary constriction will be more suitable for analysis and recognition and can be filtered automatically through iris segmentation algorithm.

As shown in Fig. 5.8, when the subjects watch the capture window, the checkerboard stimulus can be seen through the lateral prism, because there is one specified metal coating layer on lateral prism's hypotenuse surface, which divides the light ray into two parts: one is NIR light (wavelength longer than 650 nm) reflectance, and the other is the visible light (wavelength shorter than 650 nm) transmittance. NIR light is reflected by the central prism and received by camera to generate the iris images, while the visible light transmits the lateral prism between eyes and checkerboard stimulus.

The infrared distance measuring sensor is GP2D12 of Sharp, with integrated signal processing and analog voltage output. The measuring distance range is from 100 to 800 mm. When the infrared distance measuring sensor gets the distance between the subject and the capture unit, both the distance indicator and the speaker provide subject guidance so the correct capture position is achieved. According to the focus range of AF lens (from 160 to 200 mm), if the distance is out of the above range, the distance indicator will blink with red or blue LEDs and the voice instructions will be "Please move back" or "Please move closer."

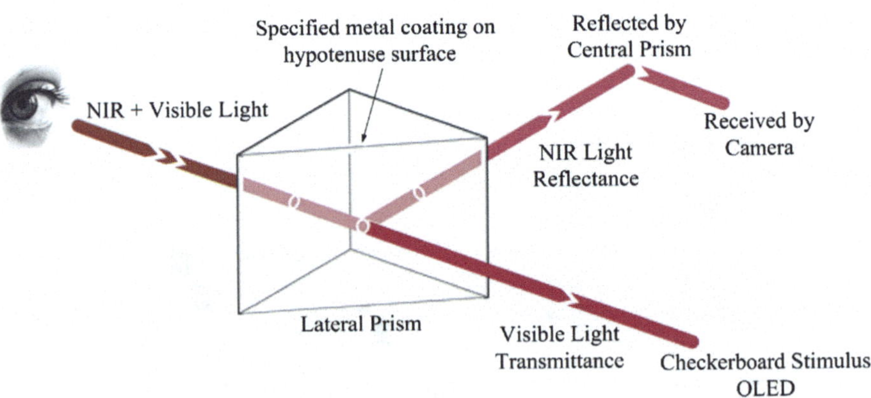

Fig. 5.8 Optical path of lateral triangular prism

In effect, subjects can make quick adjustment on their poses according to the guidance information, including light and voice. When the capture is completed, the distance indicator will automatically blink with green LEDs to tell the subject that it is successful in capturing iris images with good quality.

5.2.4 *Control Unit*

The control unit is a TI ARM Cortex-M3 Stellaris LM3S9B92 main-board, running uC/OS. The control unit can communicate with iris recognition server via USB 2.0 interface, transmit the iris image data, drive AF lens, and synchronize the other three units: capture, illumination, and interaction.

It is desirable to capture a sequence of iris images with the same or similar occlusion, pupil dilation and focusing accuracy in a short time, to introduce as little interference to multispectral fusion and recognition as possible. So, we use AF mode of lens and max frame rate of CMOS, and capture continuously a sequence of iris images under different wavelengths illumination switched automatically with high speed.

There are three factors influencing capture speed: the switching speed of wavelengths, the frame rate of CMOS, and the focusing speed of lens. The wavelengths switching time is small enough to be negligible, and the frame rate is limited by the bandwidth of CMOS, almost no room for improvement. So, the focusing speed is the most important factor to be improved. The refractive index of the same lens changed with the spectral wavelength, so the lens focus should be adjusted each time that one wavelength illumination was switched on. Among the 3 wavelengths, 700 nm is the shortest, with the largest refractive index and minimum object distance, and 850 nm is the longest, with the smallest refractive index and maximum object distance, and 780 nm is between them. The lens focus in traditional capture device is manually adjusted by operator, and it is impossible to meet the requirement of high-speed capture, so AF lens is necessary in this design.

The focusing speed is limited by two factors: the efficiency of focusing algorithm and the mechanical movement of lens. The latter almost cannot be changed, while the former has much room for improvement. We use Daugman's convolution matrix method (Daugman 2004) as the focusing algorithm. The recognition server is running the focusing algorithm to estimate the focus quality of a multispectral image and to indicate the direction of lens focus's movement for higher focus score, until the focus score reaches a preset threshold, then the lens is correctly focused. The lens driving signal is transferred from server to ARM board via USB interface, and then to New Scale M3-F Focus Module via I^2C interface. The computational time of focusing algorithm on each image is less than 10 ms on computer with Intel Pentium Processor G620 at 2.60 GHz. Including the mechanical movement of lens, the action cycle of one focusing will take not more than 300 ms.

The control unit manages the working process of one multispectral data collection cycle and synchronizes the other three units: capture, illumination, and interaction, as shown in Fig. 5.9. Once subject moves closer to the capture device, the distance measuring sensor will output the distance voltage signal. The work cycle of distance measuring sensors is 40 ms. When the ARM main-board receives voltage signal and serializes distance value, a drive signal will be sent to the distance guide indicator, instructing subject to move closer or back. At the same time, the checkerboard stimulus OLED starts to flicker between two reversal patterns. If subject moves into the focus range of AF lens, the illumination unit will be turned on in order of the wavelengths: 700, 780, and 850 nm. At every time of one wavelengths illumination turned on, the lens begins auto-focusing and CMOS begins acquisition. Taking into account that it will take some time for subject to move into the focus rage, a complete capture cycle generally can be completed within 2 or 3 s, and then, the images of two irises will be transferred to server via USB 2.0 interface.

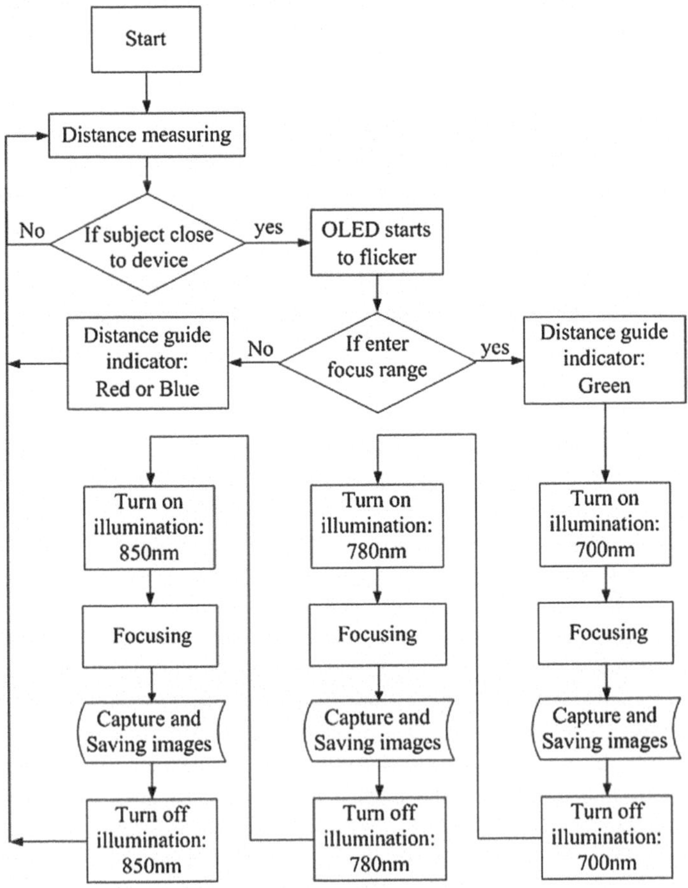

Fig. 5.9 Block diagram of multispectral iris acquisition process

5.3 Multispectral Image Fusion

5.3.1 *Proposed Iris Image Capture Device*

In this section, a series of experiments is performed to evaluate the performance of the proposed multispectral iris capture device. First, a multispectral iris image database is created by the proposed capture device. Then, we use the iris score-level fusion to further investigate the effectiveness of the proposed capture device by the 1-D Log-Gabor wavelet filter approach proposed by Masek (Masek 2003).

According to the proposed design, we developed a multispectral iris image capture device, which is shown in Fig. 5.10a, b. The dimension of this device is 162 mm (width) × 51 mm (height) × 61 mm (thickness). The device's working distance is about 160–200 mm. Through a USB 2.0 cable, this device can connect to the server computer running recognition algorithm and complete a full process of multispectral online iris identification.

Before image acquisition, Micron CMOS camera is switched to autoexposure mode and shutter time is preset as 1/30 s, in order to minimize the difference in the brightness of images and to adapt to the max frame rate 30 fps of CMOS. For each subject, the capture device will automatically start a multispectral data collection cycle once it detects that the subject has move into the focus range. In acquisition process, the subjects do not need to do other things except to watch the capture window. Each wavelength illuminator is switched on according to the preset order automatically, and two images corresponding to the left and right eye are captured and transferred to the server, then the current wavelength illuminator is turned off and the next wavelength illuminator repeats the above process. The scene of the multispectral data collection is shown in Fig. 5.11. The operator held the handle of the proposed handheld capture device, which is aiming at the user's eyes, and moved it forward or backward according to the voice prompts. A complete multispectral dual-iris recognition cycle (capture, fusion, and recognition of six images from left and right iris under three different wavelength bands) can be completed within 2 or 3 s. Using Intel atom N270 CPU (single core) running at 1.6 GHz clock rate, which is the low-end product and has the lowest computing performance

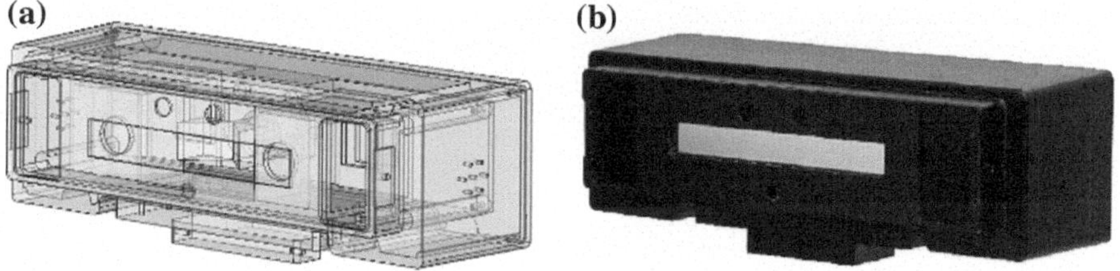

Fig. 5.10 Multispectral capture device. **a** Internal structure, **b** external form

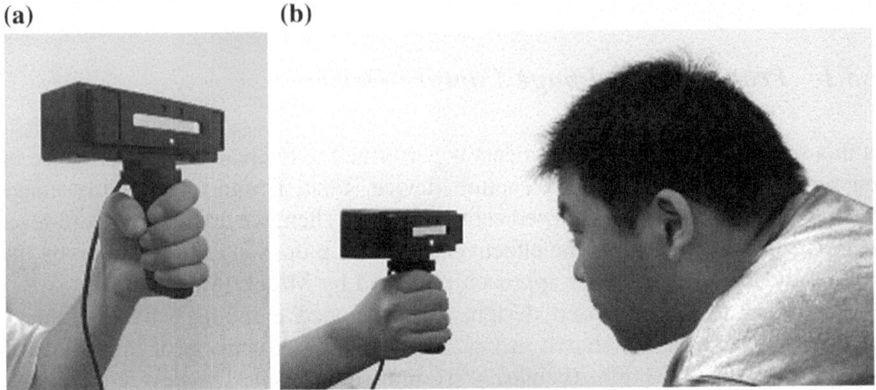

Fig. 5.11 **a** Device with handle, **b** the scene of multispectral iris capture

among all Intel desktop CPU, the Iris template extraction time for one 1600 × 1200 pixels iris image is 0.13–0.15 s. Using Neurotechnology Company's VeriEye SDK on Intel Core 2 Q9400 CPU (4 cores) running at 2.67 GHz clock rate, which is nearly highest computing performance among all Intel desktop CPU, the Iris template extraction time for one 640 × 480 pixels iris image is 0.11–0.13 s (VeriEye SDK 2012), almost equal with the our time consumption on Intel N270. Compared with other major commercial iris recognition algorithms, our recognition system has the obvious speed advantage. For the operation demonstration video, please download from the following link: http://www.youtube.com/watch?v=3xfLb_NLEmM&feature=plcp.

The multispectral example iris images that are captured by the proposed device are shown in Fig. 5.12. As shown in Fig. 5.12a–c, we can know that the captured iris image quality is very good: the focus is accurate, and the pupil radius has a relatively consistent and small size. The glittering is controlled in the pupil area in most cases, so it does not affect the iris segmentation and recognition.

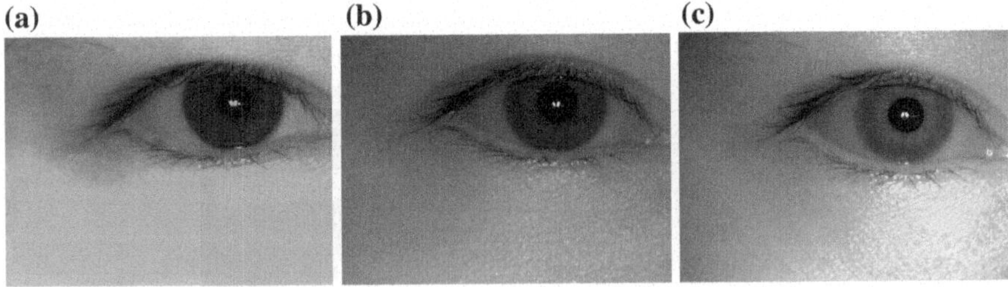

Fig. 5.12 The multispectral ample iris images: **a** captured under 700 nm, **b** captured under 780 nm, **c** captured under 850 nm, from one multispectral data collection cycle for the same iris

5.3.2 *Iris Database*

A dataset contained samples from 100 irises was used to conduct the following study. Our database, which used the proposed multispectral device, was created with 50 subjects. In this dataset, 35 are male. The age distribution is as follows: younger than 30 years comprise 84 %, and between 30 and 40 years comprise about 10 %, see Table 5.2.

The resolution of one pair of iris images (one pair including iris image captured from left and right eye) is 1600 × 1200, and the distance between the device and the subject is about 180 mm. Ten pairs of images which were taken from the same subject are selected under each of three spectral bands corresponding to 700, 780, and 850 nm, respectively. In total, we collected 1500 pairs of iris images (3000 images) for this database, which are used as a unique session in experiment.

5.3.3 *Score Fusion and Recognition*

We can locate and segment pupil in iris image based on the gray level threshold method. After segmentation, we can use the homogenous rubber sheet model devised by Daugman (1993) to normalize the iris image, by remapping each pixel within the iris region to a pair of polar coordinates (r, θ) where r is on the interval [0, 1] and θ is angle $[0, 2\pi]$. So, we can obtain six normalized patterns from one data collection cycle.

1-D Log-Gabor wavelet recognition method proposed by Masek (Masek 2003) is used for encoding in our experiments. 1-D Log-Gabor band pass filter can be efficient in angular feature extraction, which is the most distinctive and stable texture information, and ignore the radial feature extraction, which is easily interfered by dilated pupil. 1-D Log-Gabor wavelet method is the most popular comparison method used in the literature due to the accessibility of their source code. We use the performance of 1-D Log-Gabor wavelet method as the benchmark, to access the improvement of recognition performance after multispectral iris image fusion.

1-D Log-Gabor wavelet feature encoding method proposed by Masek is implemented, by convolving the normalized pattern with 1-D Log-Gabor wavelets. The rows of the 2-D normalized pattern (the angular sampling lines) are taken as the

Table 5.2 The composition of the multispectral iris image dataset

Total number of subjects	50
Number of males	35
Number of females	15
Age 21–30	42
Age 31–40	5
Age 41–50	3

1-D signal, and each row corresponds to a circular ring on the iris region. The angular direction is taken rather than the radial one, which corresponds to columns of the normalized pattern, since the iris maximum independence occurs in the angular feature extraction, which is the most distinctive and stable texture information, and ignores the radial feature extraction, which is easily interfered by dilated pupil. We revisit the default parameter values used by Masek, to get higher recognition performance (Peters 2009). All parameters revised in this work are as follows: angular resolution and radial resolution of normalized image, center wavelength and filter bandwidth of 1-D Log-Gabor filter, and fragile bit percentage, as in Table 5.3. After we apply the 1-D Log-Gabor wavelet filter to the normalized iris image, we quantize the result to create the iris code bit vectors by determining the quadrant of the response in the complex plane. This gives us an iris code bit vectors that have twice as many bits as the normalized iris image had pixels.

In our work, two single-eye iris code bit vectors $\text{code}_{\text{left}}$, $\text{code}_{\text{right}}$ generated from one pair of iris images (including left and right eye) captured under the same wavelength are combined into one dual-eye iris code bit vector $\text{code}_{\text{pair}}$ as follows:

$$\text{code}_{\text{pair}} = \left[\text{code}_{\text{left}};\ \text{code}_{\text{right}}\right] \tag{5.1}$$

Inspired by the matching scheme of Daugman (2007), the binary Hamming distance is used and the similarity between two iris images is calculated by using the exclusive-OR operation. Under wavelength i, based on two pairs whose two iris code bit vectors are denoted as $\left\{\text{code}_{\text{pair}\,A,i},\ \text{code}_{\text{pair}\,B,i}\right\}$ and mask bit vectors denoted as $\left\{\text{mask}_{\text{pair}\,A,i},\ \text{mask}_{\text{pair}\,B,i}\right\}$, we can compute the raw Hamming distance HDraw as follows:

$$\text{HD}_{\text{raw},i}\ (A, B) = \frac{\left\|\left(\text{code}_{\text{pair}\,A,i} \otimes \text{code}_{\text{pair}\,B,i}\right)\cap\text{mask}_{\text{pair}\,A,i}\cap\text{mask}_{\text{pair}\,B,i}\right\|}{\left\|\text{mask}_{\text{pair}\,A,i}\cap\text{mask}_{\text{pair}\,B,i}\right\|} \tag{5.2}$$

Suppose there are k bands of wavelengths, corresponding to k kinds of iris code bit vectors $\left(\text{code}_{\text{pair}\,A,i},\ i = \{1, 2...k\}\right)$ and k kinds of mask bit vectors $\left(\text{mask}_{\text{pair}\,A,i},\ i = \{1, 2...k\}\right)$ for one pair of irises. For two pairs of irises from different subjects A and B, the distance using simple sum rule is defined as:

Table 5.3 Listing of parameters revised for multispectral iris images with the initial value from Masek

Parameter	Initial Value	Revised Value
Angular resolution(0)	240 pixels	360 pixels
Radial resolution (r)	20 bands	60 bands
Center wavelength (λ)	13 pixels	16 pixels
Filter bandwidth (σ/f)	0.5	0.4
Row averaging	NA	3 rows
Fragile bit percentage	NA	Yes

$$\mathrm{HD}_{\mathrm{sum}}(A, B) = \sum_{i=1}^{k} \mathrm{HD}_{\mathrm{raw},i}(A, B) \quad (5.3)$$

Generally, more texture information across all different wavelengths is used; better recognition performance could be achieved. However, since maybe there is some overlapping of the discriminating information between different bands of wavelengths, simple sum of the matching scores of all bands may not improve much the final accuracy. The overlapping part between the two iris code bit vectors under two different wavelengths will be counted twice by using the sum rule (Eq. 5.3). Such kind of over-computing may make the simple score-level fusion fail.

When a score-level fusion strategy could reduce the overlapping effect, better verification results can be expected. In combinatorics, the inclusion–exclusion principle (Comtet 1974) (which is attributed to Abraham de Moivre) is an equation relating the sizes of two sets and their union. For finite sets $A_1, \ldots, A_n$, one has the identity as follows:

$$\left|\bigcup_{i=1}^{n} A_i\right| = \sum_{i=1}^{n} |A_i| - \sum_{i,j:1 \le i \le j \le n} |A_i \cap A_j| + \sum_{i,j,k:1 \le i \le j \le k \le n} |A_i \cap A_j \cap A_k| - \ldots + (-1)^{n-1} |A_1 \cap \ldots \cap A_n| \quad (5.4)$$

where $\cup$ denotes union, and $\cap$ denotes intersection. So, based on Eq. 5.4, a score-level fusion rule is defined which tends to minimize the overlapping effect on the fused score as follows:

$$\mathrm{HD}_{\mathrm{sum}(1,2)}(A, B) = HD_{\mathrm{raw},1}(A, B) + \mathrm{HD}_{\mathrm{raw},2}(A, B) - \frac{\mathrm{HD}_{\mathrm{raw},1}(A, B) + \mathrm{HD}_{\mathrm{raw},2}(A, B)}{2} \times P_{1,2}(A, B) \quad (5.5)$$

where $P_{1,2}(A, B)$ is the overlapping percentage between two iris code bit vectors exacted from the same iris but captured under two different bands of wavelength, defined as follows:

$$P_{1,2}(A, B) = 1 - \left(\frac{\left\| \left(\mathrm{code}_{\mathrm{pair}\,A,1} \otimes \mathrm{code}_{\mathrm{pair}\,A,2}\right) \cap \mathrm{mask}_{\mathrm{pair}\,A,1} \cap \mathrm{mask}_{\mathrm{pair}\,A,2} \right\|}{\left\| \mathrm{mask}_{\mathrm{pair}\,A,1} \cap \mathrm{mask}_{\mathrm{pair}\,A,2} \right\|} + \frac{\left\| \left(\mathrm{code}_{\mathrm{pair}\,B,1} \otimes \mathrm{code}_{\mathrm{pair}\,B,2}\right) \cap \mathrm{mask}_{\mathrm{pair}\,B,1} \cap \mathrm{mask}_{\mathrm{pair}\,B,2} \right\|}{\left\| \mathrm{mask}_{\mathrm{pair}\,B,1} \cap \mathrm{mask}_{\mathrm{pair}\,B,2} \right\|} \right) \times \frac{1}{2} \quad (5.6)$$

Similarly, we could extend the multispectral score fusion scheme (Gong et al. 2012a, b) to fuse more bands of wavelength, e.g., 3 spectral bands as in Eq. 5.7:

$$\begin{aligned}
\mathrm{HD}_{\mathrm{sum}(1,2,3)}(A,B) = {} & \mathrm{HD}_{\mathrm{raw},1}(A,B) + \mathrm{HD}_{\mathrm{raw},2}(A,B) + \mathrm{HD}_{\mathrm{raw},3}(A,B) \\
& - \frac{\mathrm{HD}_{\mathrm{raw},1}(A,B) + \mathrm{HD}_{\mathrm{raw},2}(A,B)}{2} \times P_{1,2}(A,B) \\
& - \frac{\mathrm{HD}_{\mathrm{raw},1}(A,B) + \mathrm{HD}_{\mathrm{raw},3}(A,B)}{2} \times P_{1,3}(A,B) \\
& - \frac{\mathrm{HD}_{\mathrm{raw},2}(A,B) + \mathrm{HD}_{\mathrm{raw},3}(A,B)}{2} \times P_{2,3}(A,B) \\
& + \frac{\mathrm{HD}_{\mathrm{raw},1}(A,B) + \mathrm{HD}_{\mathrm{raw},2}(A,B) + \mathrm{HD}_{\mathrm{raw},3}(A,B)}{3} \times P_{1,2,3}(A,B)
\end{aligned} \tag{5.7}$$

We usually evaluate recognition accuracy according to three indicators: FAR (false acceptance rate, a measure of the likelihood that the access system will wrongly accept an access attempt), FRR (false rejection rate, the percentage of identification instances in which false rejection occur), and EER (equal error rate, the value where FAR and FRR are equal). So, we calculate the FAR, FRR, and EER based on the images of single wavelength: 700, 780, 850 nm and the images of multispectral fusion. The intra-spectral genuine and intra-spectral impostor scores are used to compute the EER and FRR.

We capture 500 pairs of iris images (1000 iris images) under each wavelength. So in total, there are 1500 pairs of iris images (3000 iris images) across all three wavelengths. Under each wavelength, 2,250 intra-spectral genuine scores and 122,500 intra-spectral impostor scores were generated from the dual-eye iris code ($\mathrm{code}_{\mathrm{pair}}$) matching, and 4,500 intra-spectral genuine scores and 245,000 intra-spectral impostor scores were generated from the single-eye iris code (($\mathrm{code}_{\mathrm{left}}$ or $\mathrm{code}_{\mathrm{right}}$) matching.

In both dual-eye and single-eye iris recognition, sum score-level fusion (Gong et al. 2012a, b) is used as the fusion technique on the dataset, to generate the multispectral genuine scores and imposter scores based on all three wavelengths. The normalized histogram plots of $\mathrm{HD}_{\mathrm{norm}}$ for the dual-eye images of three wavelengths and multispectral fusion are shown in Fig. 5.13. The normalized histogram plots of HDnorm for the single-eye images of three wavelengths and multispectral fusion are shown in Fig. 5.14.

5.3.4 Experimental Results and Analysis

In Figs. 5.13 and 5.14, each of four images is composed of two parts: the lower is the matching score distribution, and the upper is the magnified distribution near the cross-point of genuine curve (the left blue one) and impostor curve (the right red one). The blue curve which is to the right of the cross-point means FRR, while the

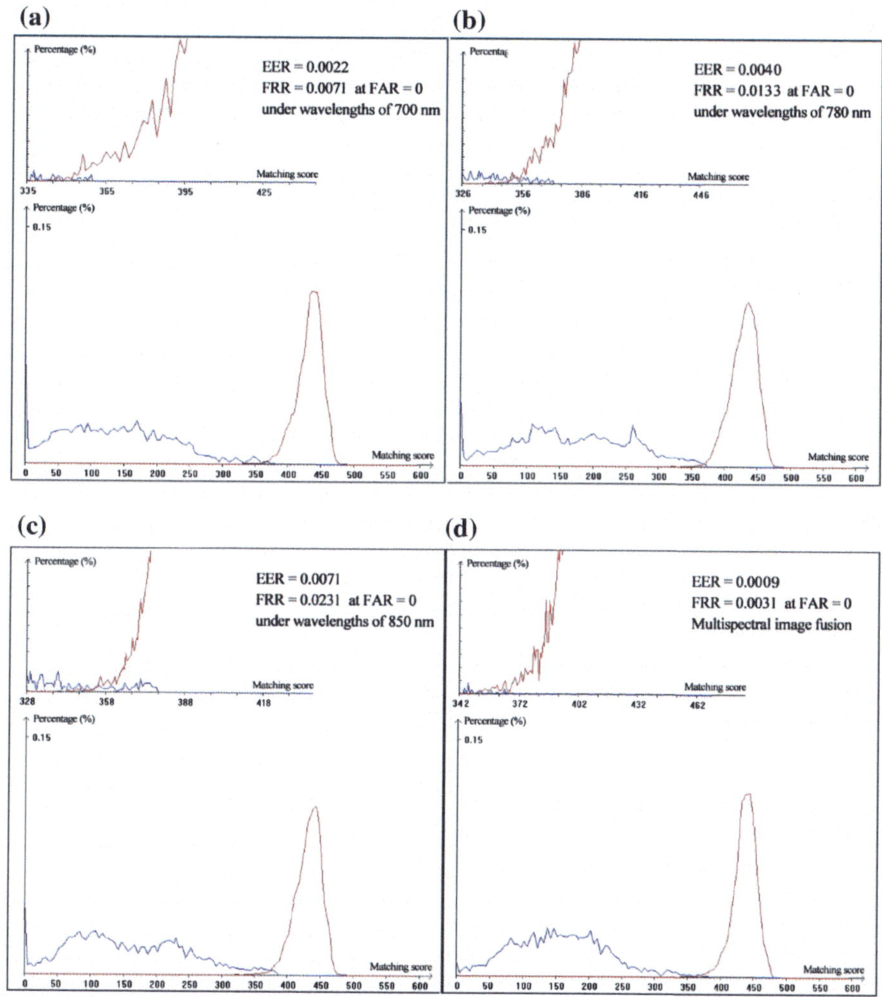

Fig. 5.13 EER and FRR (FAR = 0) based on the dual-eye images of three wavelengths and multispectral fusion: **a** 700, **b** 780 nm, **c** 850 nm, **d** multispectral score-level fusion

red curve which is to the left of the cross-point means FAR. (a)–(d) show the same characteristics: The intra-spectral genuine scores of different wavelengths have similar median values and models and are mostly spread around the corresponding median value. The intra-spectral impostor scores are observed to be fairly well separated from the intra-spectral genuine scores.

The EERs (where FRR = FAR) and FRRs (where the highest degree of FAR accuracy can be obtained by the user, FAR = 0) based on the images of single wavelength and the images of multispectral fusion can be clearly compared. See Figs. 5.13 and 5.14, in both dual-eye and single-eye matching, the EER (where

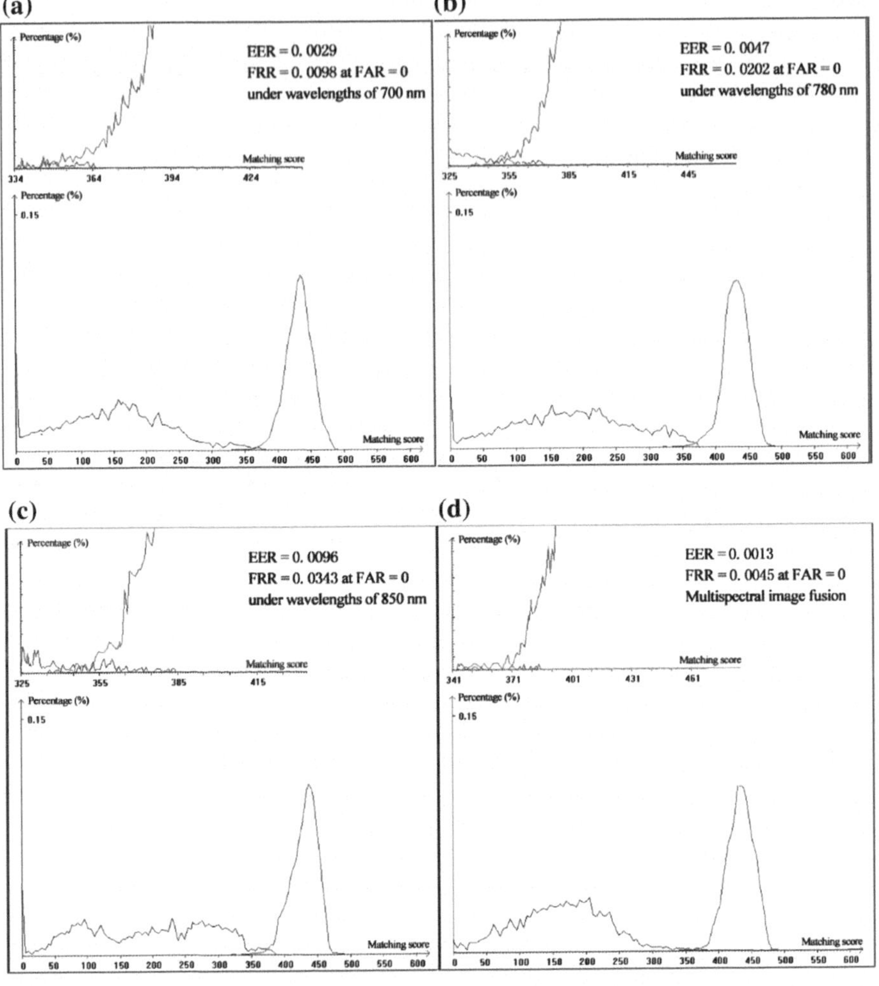

Fig. 5.14 EER and FRR (FAR = 0) based on the single-eye images of three wavelengths and multispectral fusion: **a** 700, **b** 780 nm, **c** 850 nm, **d** multispectral score-level fusion

FRR = FAR) and FRR (where FAR = 0) reach their lowest values based on multispectral score-level fusion, while the ones based on the images of single wavelength are relatively higher, which means the best recognition performance is achieved by multispectral fusion as in see Table 5.4.

In dual-eye matching, according to the FRR (where FAR = 0), 1-D Log-Gabor wavelet recognition method with multispectral score-level fusion is 87 % lower than the one under 850 nm without multispectral image fusion, 77 % lower than the one under 780 nm without multispectral image fusion, 56 % lower than the one under 700 nm without multispectral image fusion. According to the EER (where FRR = FAR), 1-D Log-Gabor wavelet recognition method with multispectral image

Table 5.4 The comparison of EER and FRR between dual-eye and single-eye matching

Wavelengths (nm)	Dual-eye iris recognition		Single-eye iris recognition	
	EER (FAR = FRR)	FRR (FAR = 0)	EER (FAR = FRR)	FRR (FAR = 0)
700	0.0022	0.0071	0.0029	0.0098
780	0.004	0.0133	0.0047	0.0202
850	0.0071	0.0231	0.0096	0.0343
Multispectral score-level fusion	0.0009	0.0031	0.0012	0.0045

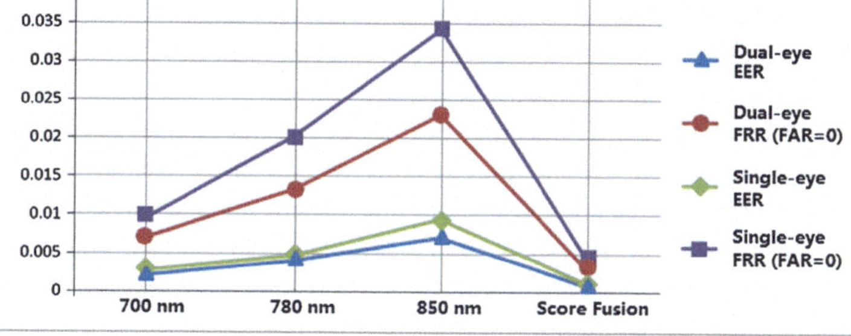

fusion is 87 % lower than the one under 850 nm without multispectral image fusion, 78 % lower than the one under 780 nm without multispectral image fusion, and 59 % lower than the one under 700 nm without multispectral image fusion.

In single-eye matching, according to the FRR (where FAR = 0), 1-D Log-Gabor wavelet recognition method with multispectral score-level fusion is 87 % lower than the one under 850 nm without multispectral image fusion, 78 % lower than the one under 780 nm without multispectral image fusion, and 54 % lower than the one under 700 nm without multispectral image fusion. According to the EER (where FRR = FAR), 1-D Log-Gabor wavelet recognition method with multispectral image fusion is 86 % lower than the one under 850 nm without multispectral image fusion, 72 % lower than the one under 780 nm without multispectral image fusion, and 55 % lower than the one under 700 nm without multispectral image fusion.

According to the FRR (where FAR = 0) and EER (where FRR = FAR), 1-D Log-Gabor wavelet recognition method in dual-eye matching is 31 % lower than the one in single-eye matching.

Up to now, no other academic organizations and the commercial companies have announced the experimental results about the multispectral iris recognition performance based on three bands of wavelengths image fusion. So, we compared our multispectral iris recognition performance with Masek's performance on iris images of single wavelength band, which is proposed by Masek (Masek 2003). Two recognition performances are based on the same template extraction algorithm, 1-D Log-Gabor wavelet filter approach. The difference is: the former is based on multispectral images and multispectral score-level fusion, and the latter is based on the images of single wavelength band.

According to the FRR (where FAR = 0), Masek got 0.0458 on LEI-a iris data set and 0.05181 on CASIA-a iris data set. Our proposed dual-eye multispectral recognition performance is about 1/10 of the above rates, showing a very significant performance advantage over traditional system.

From the above comparison, we can learn that in both dual-eye and single-eye matching, the higher iris recognition accuracy can be achieved based on the multispectral images captured by the proposed device and the score-level fusion. The simultaneous acquisition and recognition of dual-eye multispectral iris images have greater performance advantages than traditional single-eye system, in image acquisition efficiency, recognition accuracy, and adaptability to complex situations. Further, we can get two inferences, which are all innovative: First, the dual-eye multispectral iris images captured by the proposed handheld device are good enough for score-level fusion and recognition, so the proposed device meets the design requirements and can be used for dual-eye multispectral iris capture and recognition, which is the initiative in the research of multispectral iris recognition. Second, the integrated dual-eye multispectral iris recognition system consisting of the image acquisition device and the recognition server is feasible and can complete a full process of multispectral online iris identification, which is the first successful attempt of practical application of the dual-eye iris recognition based on single low-resolution camera.

5.4 Summary

A handheld system for dual-eye multispectral iris capture based on the improved "one-camera" design has been proposed. Using a set of refraction lens, we enlarged the effective resolution of 2 million pixels camera, which can cover the width of the two eyes within 1600 × 1200 pixels, capture two iris images with high resolution and high image quality, and meet the iris recognition standards ISO/IEC 2004 (200 pixels across the iris—the minimum acceptable resolution for level 2 iris features). Using this system, a complete capture cycle (including six images from left and right iris under three different wavelength bands) can be completed within 2 or 3 s, much faster than previous devices.

The system consists of the following four parts: (1) capture unit; (2) illumination unit; (3) interaction unit; and (4) control unit. It uses a low-resolution (2 million pixels) Micron CMOS camera. The matrix-arrayed LEDs across three different wavelengths (including visible light and near-infrared light) are used as the multispectral illumination unit. We design an interaction unit with the checkerboard stimulus evoking the pupillary constriction to eliminate the interference of pupil dilation. The control unit synchronizes the previous three units and makes it capturing with high speed, easy to use, and nonintrusive for users.

A series of experiments is performed to evaluate the performance of the proposed multispectral iris capture device. A multispectral iris image database is created by the proposed capture device, and then, we use the iris score-level fusion to further investigate the effectiveness of the proposed capture device by the 1-D Log-Gabor wavelet filter approach. Experimental results have illustrated the encouraging performance of the improved "one-camera" design.

In summary, the proposed handheld dual-eye capture device benefits from simplicity and lower cost, because of reduced optics, sensors, and computational needs. This system is designed to have good performance at a reasonable price so that it becomes suitable for civilian personal identification applications. For further improvement of the system, we will focus on the following issues: conducting multispectral fusion and recognition experiments on a large number of iris databases in various environments and presenting a more critical analysis of the accuracy and performance measurement of the proposed system.

References

Biom Technol (2005) Iris recognition in focus. Today 13(2): 9–11

Burge MJ, Monaco MK (2009) Multispectral iris fusion for enhancement, interoperability, and cross wavelength matching. In: SPIE Defense, Security, and Sensing: 73341D–73341D

CASIA Iris Image Database (2005) http://www.cbsr.ia.ac.cn/IrisDatabase.htm

Comtet L (1974) Advanced combinatorics: the art of finite and infinite expansions. Reidel Dordrecht: 176–177

Daugman JG (1993) High confidence visual recognition of persons by a test of statistical independence. IEEE Trans Pattern Anal Mach Intell 15(11):1148–1161

Daugman J (2004) How iris recognition works. IEEE Trans Circ Syst Video Technol 14(1):21–30

Daugman J (2007) New methods in iris recognition. IEEE Trans Syst Man Cybern B Cybern 37(5):1167–1175

Dual Iris Scanner (2014) http://64.233.183.163/url?q=http://solutions.3m.com/wps/portal/3M/en_US/Security/Identity_Management/Products_Services/Reader_Scanner_Solutions/Scanning_Solutions/Dual_Iris_Scanner/&sa=U&ei=IxCJVL_-LdeeyASxrYHoDA&ved=0CBcQFjAA&usg=AFQjCNGEYGWgyS3mATyxhw4eCnd7JocS6g

Gong Y, Zhang D, Shi P, Yan J (2012a) High-speed multispectral iris capture system design. IEEE Trans Instrum Meas 61(7):1966–1978

Gong Y, Zhang D, Shi P, Yan J (2012b) Optimal wavelength band clustering for multispectral iris recognition. Appl Opt 51(19):4275–4284

He X, Yan J, Chen G, Shi P (2008) Contactless autofeedback iris capture design. IEEE Trans Instrum Meas 57(7):1369–1375

Hollingsworth K, Bowyer KW, Flynn PJ (2009) Pupil dilation degrades iris biometric performance. Comput Vis Image Underst 113(1):150–157

IriMagic™ Series (2000) http://www.iritech.com/products/hardware/irimagic™-series

Masek L (2003) Recognition of human iris patterns for biometric identification. Doctoral dissertation, Master's thesis, University of Western Australia

Matey JR, Naroditsky O, Hanna K, Kolczynski R, LoIacono DJ, Mangru S, Zhao WY (2006) Iris on the move: acquisition of images for iris recognition in less constrained environments. Proc IEEE 94(11):1936–1947

Mobile Dual Iris Capture Device (2014) http://www.crossmatch.com/i-scan-2/

Ngo HT, Ives RW, Matey JR, Dormo J, Rhoads M, Choi D (2009) Design and implementation of a multispectral iris capture system. In: Signals systems and computers 2009 conference record of the forty-third asilomar conference, pp 380–384

Park KR, Kim J (2005) A real-time focusing algorithm for iris recognition camera. IEEE Trans Syst Man Cybern Part C Appl Rev 35(3):441–444

Peters TH (2009) Effects of segmentation routine and acquisition environment on iris recognition. Doctoral dissertation, University of Notre Dame

Ross A, Pasula R, Hornak L (2006) Exploring multispectral iris recognition beyond 900 nm. In: Proceedings of the 2006 conference on computer vision and pattern recognition workshop, 51

Shi P, Xing L, Gong Y (2003) A quality evaluation method of iris recognition system. Chin Pat 1474:345

Sun FC, Chen LY, Zhao XZ (1998) Pupillary responses evoked by spatial patterns. Acta Physiol Sinica 50(1):67–74

Tan T, Zhu Y, Wang Y (1999) Iris image capture device. Chin Pat 2392:219

VeriEye SDK (2012) http://download.neurotechnology.com/VeriEye_SDK_Brochure_2012-04-23.pdf

Vilaseca M, Mercadal R, Pujol J, Arjona M, de Lasarte M, Huertas R, Imai FH (2008) Characterization of the human iris spectral reflectance with a multispectral imaging system. Appl Opt 47(30):5622–5630

VistaEY2H Handheld Dual Iris Camera (1998) http://www.neurotechnology.com/eye-iris-scanner-vistaey2h.html

Wildes RP (1997) Iris recognition: an emerging biometric technology. Proc IEEE 85(9):1348–1363

Wilkerson CL, Syed NA, Fisher MR, Robinson NL, Albert DM (1996) Melanocytes and iris color: light microscopic findings. Arch Ophthalmol 114(4):437–442

Part III
Multispectral Palmprint Recognition

Chapter 6
An Online System of Multispectral Palmprint Verification

Abstract Palmprint is a unique and reliable biometric characteristic with high usability. With the increasing demand of highly accurate and robust palmprint authentication system, multispectral imaging has been employed to acquire more discriminative information and increase the anti-spoof capability of palmprint. This chapter presents an online multispectral palmprint system that could meet the requirement of real-time application. A data acquisition device is designed to capture the palmprint images under blue, green, red, and near-infrared (NIR) illuminations in less than 1 s. A large multispectral palmprint database is then established to investigate the recognition performance of each spectral band. Our experimental results show that the red channel achieves the best result, while the blue and green channels have comparable performance but are slightly inferior to the NIR channel. After analyzing the extracted features from different bands, we propose a score-level fusion scheme to integrate the multispectral information. The palmprint verification experiments demonstrated the superiority of multispectral fusion to each single spectrum, which results in both higher verification accuracy and anti-spoofing capability.

Keywords Palmprint verification · Biometrics · Multispectral · Score-level fusion · Orientation code

6.1 Introduction

Biometrics refers to the study of methods for recognizing humans based on one or more physical or behavioral traits (Jain et al. 1999; Zhang 2000). As an important biometrics characteristic, palmprint has been attracting much attention (Wu et al. 2003) because it has merits such as high speed, user-friendliness, low cost, and high accuracy. However, there is room for improvement of online palmprint systems in the aspects of accuracy and capability of spoof attacks (Schukers 2002). Although 3-D imaging could be used to address these issues, the expensive and bulky device makes it difficult to be applied for real applications. One solution to these problems

D. Zhang et al., *Multispectral Biometrics*,
DOI 10.1007/978-3-319-22485-5_6

can be multispectral imaging (Rowe et al. 2005; Singh e al. 2008; Park and Kang 2007), which captures an image in a variety of spectral bands. Each spectral band highlights specific features of the palm, making it possible to collect more information to improve the accuracy and anti-spoofing capability of palmprint systems.

Multispectral analysis has been used in palm-related authentication (Hao et al. 2007, 2008; Rowe et al. 2007; Likforman-Sulem et al. 2007; Wang et al. 2008). Rowe et al. (2007) proposed a multispectral whole-hand biometric system. The object of this system was to collect palmprint information with clear fingerprint features, and the imaging resolution was set to 500 dpi. However, the low speed of feature extraction and feature matching makes it unsuitable for real-time applications. Likforman-Sulem et al. (2007) used multispectral images in a multimodal authentication system; however, their system used an optical desktop scanner and a thermal camera which make the system very costly. The imaging resolution is also too high (600 dpi, the FBI fingerprint standard) to meet the real-time requirement in practical biometric systems. Wang et al. (2008) proposed a palmprint and palm vein fusion system, which could acquire two kinds of images simultaneously. The system uses one color camera and one near-infrared (NIR) camera and requires a registration procedure of about 9 s. Hao et al. (2007, 2008) developed a contact-free multispectral palm sensor. However, the image quality is limited and hence the recognition accuracy is not very high. Overall, multispectral palmprint scanning is a relatively new topic and the above-mentioned works stand for the state-of-the-art work.

The information presented by multiple biometric measures can be consolidated at four levels: image level, feature level, matching score level, and decision level (Ross et al. 2006). Wang et al. (2008) fused palmprint and palm vein images by using a novel edge-preserving and contrast-enhancing wavelet fusion method for use of personal recognition system. Some good results in accuracy were reported, but the image registration procedure in it takes 9 s, which hinders it from real-time implementation. Hao et al. (2007) evaluated several well-known image-level fusion schemes for multispectral palm images. However, their dataset was too small (84 images) to be conclusive. In 2008, Hao et al. extended their work to a larger database and proposed a new feature-level registration method for image fusion. The results by various image fusion methods were also improved. Although image- and feature-level fusion can integrate the information provided by each spectral band, the required registration procedure is often too time-consuming (Wang et al. 2008). As to matching score-level fusion and decision-level fusion, it has been found (Ross et al. 2006) that the former works better than the later because match scores contain more information about the input pattern and it is easy to access and combine the scores generated by different matchers. For these reasons, information fusion at score level is the most commonly used approach in multimodal biometric systems and multispectral palmprint systems (Rowe et al. 2007; Likforman-Sulem et al. 2007).

In this chapter, we propose a low-cost multispectral palmprint system that can operate in real time and acquire high-quality images. The proposed system collects palmprint images in visible and near-infrared spectra. Compared with traditional palmprint recognition approaches, the developed system can improve much the recognition accuracy by fusing the information provided by multispectral palmprint images at the score level. On the other hand, by utilizing the correlation between spectra, the system is robust to anti-spoof attack.

The rest of the chapter is organized as follows. Section 6.2 describes the developed online multispectral palmprint system. Section 6.3 introduces multispectral palmprint analysis methodology. Section 6.4 reports our experimental results in a large multispectral palmprint database. Section 6.5 gives the conclusions and future work.

6.2 The Online Multispectral Palmprint System Design

In this section, we describe the components of our proposed system and its parameters. Two basic considerations in the design of a multispectral palmprint system are the color-absorptive and color-reflective characteristics of human skin and the light spectra to be used when acquiring images. Human skin is made up of three layers: the epidermis, dermis, and subcutis as shown in Fig. 6.1. Each layer

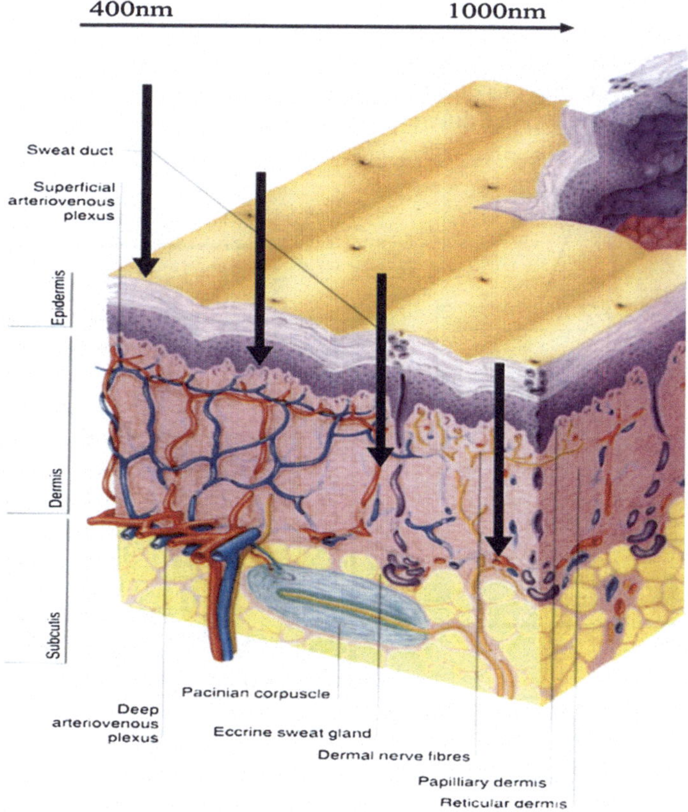

Fig. 6.1 Cross-sectional anatomy of the skin

will contain a different proportion of blood and fat. The epidermis also contains melanin, while the subcutis contains veins (Gawkrodger 2002). Different light wavelengths will penetrate to different skin layers and illuminate in different spectra. NIR light penetrates human tissue further than visible light, and blood absorbs more NIR energy than surrounding tissue (e.g., fat or melanin) (Zharov et al. 2004). Our system acquires spectral information from all the three dermal layers by using both the visible bands and NIR band. In the visible spectrum, a three mono-color light-emitting diode (LED) array is used with red peaking at 660 nm, green peaking at 525 nm, and blue peaking at 470 nm. In the NIR spectrum, an NIR LED array peaking at 880 nm is used. It has been shown that light in the 700–1000 nm range can penetrate human skin, while 880–930 nm provides a good contrast of subcutaneous veins (Zharov et al. 2004).

Figure 6.2 shows the structure of the designed multispectral palmprint image acquisition device and Fig. 6.3 shows the prototype of our device. It is composed of a charge-coupled device (CCD) camera, lens, an analog-to-digital (A/D) converter, a multispectral light source, and a light controller. A monochromatic CCD is placed at the bottom of the device. The A/D converter connects the CCD and the computer. The light controller is used to control the multispectral light.

The system can capture palmprint images in a resolution of either 352 by 288 or 704 by 576. A user is asked to put his/her palm on the platform. Several pegs serve as control points for the placement of the user's hands. Four palmprint images of the palm are collected under different spectral lights. The switching time between

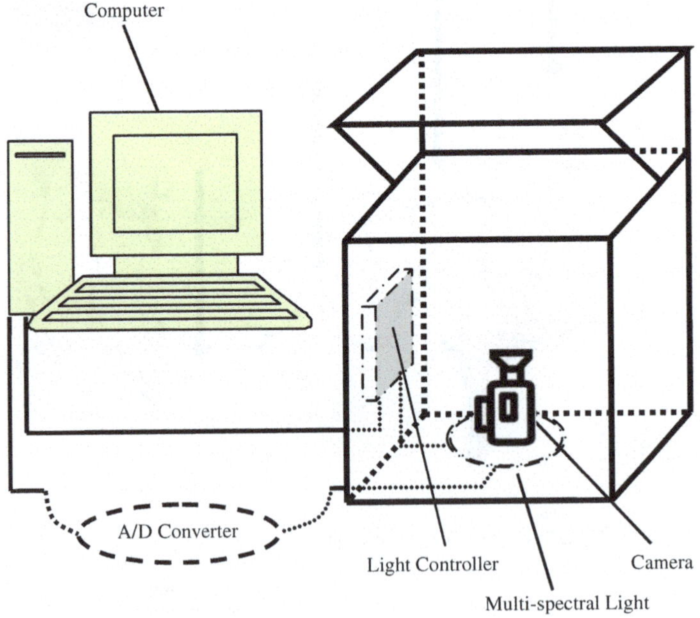

Fig. 6.2 The structure of the multispectral palmprint acquisition device

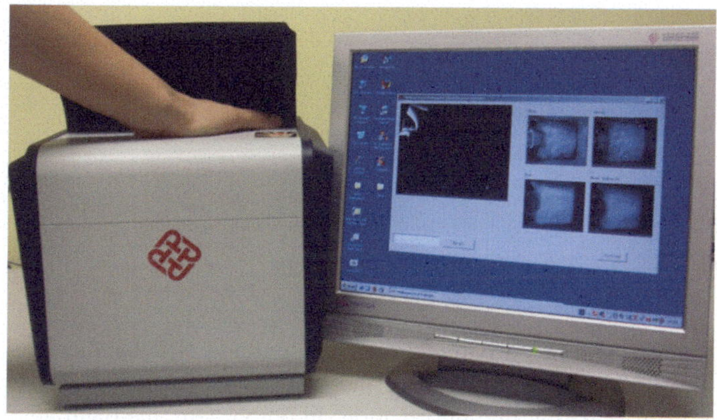

Fig. 6.3 Prototype of the proposed multispectral palmprint system

the two consecutive lights is very short and the four images can be captured in a very short time (<1 s).

Figure 6.4 shows a typical multispectral palmprint sample in the (a) blue, (b) green, (c) red, and (d) NIR bands. It can be observed that the line features are

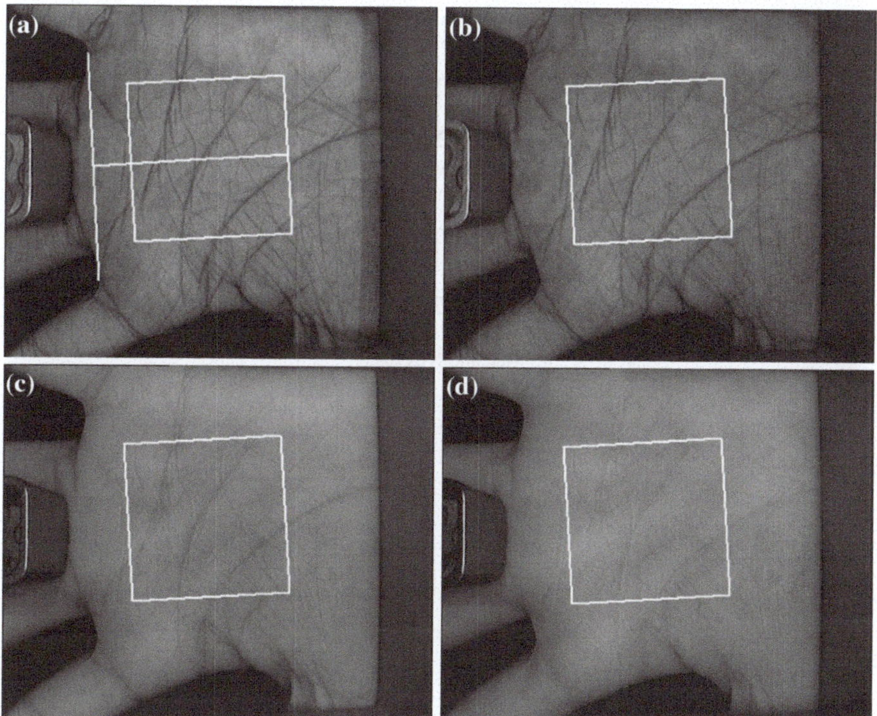

Fig. 6.4 A typical multispectral palmprint sample. **a** Blue; **b** Green; **c** Red; **d** NIR. The *white square* is the region of interest (ROI) of the image

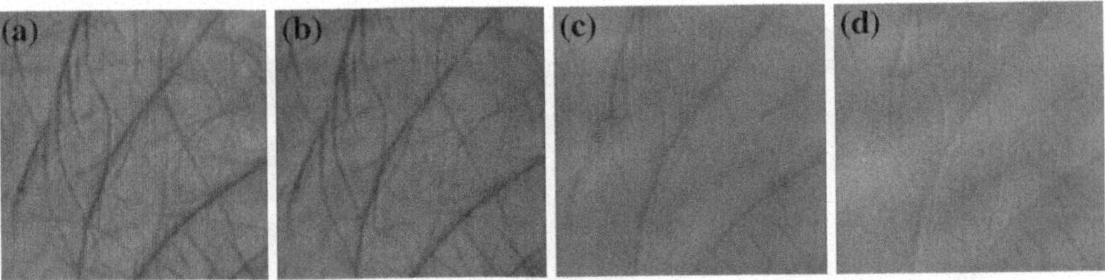

Fig. 6.5 ROI of Fig. 6.4. **a** Blue; **b** Green; **c** Red; **d** NIR

clearer in the blue and green bands than in the red and NIR bands. While the red band can reveal some vein structure, the NIR band can show the palm vein structures as well as partial line information. In our system, the palm vein structure acquired in the NIR band is not as clear as that reported in Zharov et al. (2004) because the CCD in our system is a standard closed-circuit television (CCTV) camera, instead of a near-infrared-sensitive camera, to reduce cost. Moreover, we do not add an infrared filter in front of the CCD because the infrared filter would cut

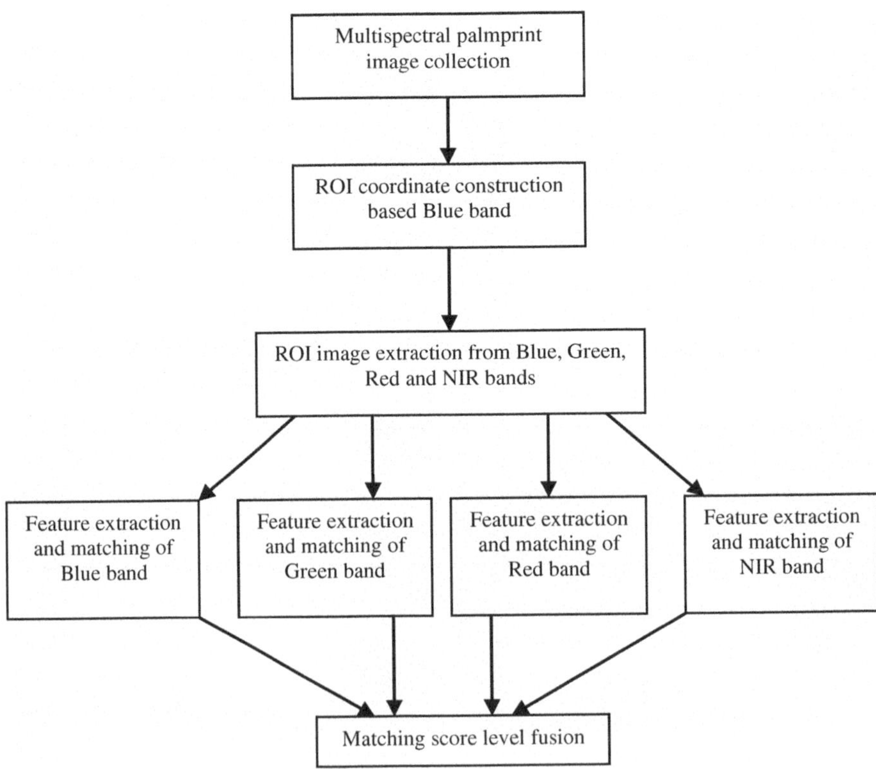

Fig. 6.6 Framework of the online verification system

off the visible light and affect the acquisition of clear palmprint images under the visible spectrum.

A region of interest (ROI) will be extracted from the palmprint image for further feature extraction and matching. This can reduce the data amount in feature extraction and matching and reduce the influence of rotation and translation of the palm. In this chapter, the ROI extraction algorithm in Zhang et al. (2003) is used and applied to the blue band to find the ROI coordinate system. After ROI extraction, the translation or rotation is usually small between two images. Thus, no more registration procedure is necessary. Figure 6.4 shows the ROI of the palmprint image, and Fig. 6.5 shows the cropped ROI images.

After obtaining the ROI for each band, feature extraction and matching will be applied to these ROI images. The final verification decision will be made by score-level fusion of all bands. Figure 6.6 shows the framework of the proposed online verification system.

6.3 Multispectral Palmprint Image Analysis

In this section, we first describe how features are extracted and matched for each spectrum band of multispectral palmprint images and then analyze the correlation between the spectral bands. Finally, a score-level fusion scheme is presented for decision making.

6.3.1 Feature Extraction and Matching for Each Band

There are mainly three kinds of algorithms currently used in palmprint authentication: subspace learning (Wu et al. 2003; Connie et al. 2005; Hu et al. 2007), line detection (Han et al. 2003; Wu et al. 2006), and texture-based coding (Zhang et al.2003; Kong et al. 2006; Kong and Zhang 2004; Jia et al. 2008). Online system usually uses texture-based coding because it has high accuracy and is robust to illumination and fast for matching. The proposed system employs orientation-based coding (Kong and Zhang 2004), a state-of-the-art texture-based coding algorithm, for feature extraction.

By viewing the line features in palmprint images as negative lines, we apply six Gabor filters along different directions ($\theta_j = j\pi/6$, where $j = \{0, 1, 2, 3, 4, 5\}$) to the palmprint images for orientation feature extraction. For each pixel, the orientation corresponding to the minimal response is taken as the feature at this pixel (Kong and Zhang 2004). The employed Gabor filter is as follows:

$$\psi(x, y, \omega, \theta) = \frac{\omega}{\sqrt{2\pi}\kappa} e^{-\frac{\omega^2}{8\kappa^2}(4x'^2 + y'^2)} \left(e^{i\omega x'} - e^{-\frac{\kappa^2}{2}} \right) \quad (6.1)$$

where $x' = (x - x_0)\cos\theta + (y - y_0)\sin\theta, y' = -(x - x_0)\sin\theta + (y - y_0)\cos\theta$ is the center of the function, ω is the radial frequency in radians per unit length, θ is the orientation of the Gabor functions in radians, $\kappa = \sqrt{2\ln 2}\left(\frac{2^\varphi + 1}{2^\varphi - 1}\right)$ and φ is the half-amplitude bandwidth of the frequency response. To reduce the influence of illumination, the DC (direct current) is removed from the filter. Figure 6.7 shows some feature maps extracted from Fig. 6.5 using a variety of parameters.

Since there are totally six different orientations, we can code them by using 3 bits as listed in Table 6.1. This coding scheme is to make the bitwise difference proportional to the angular difference (Kong and Zhang 2004). So the difference between two orientation maps could be measured by using the Hamming distance:

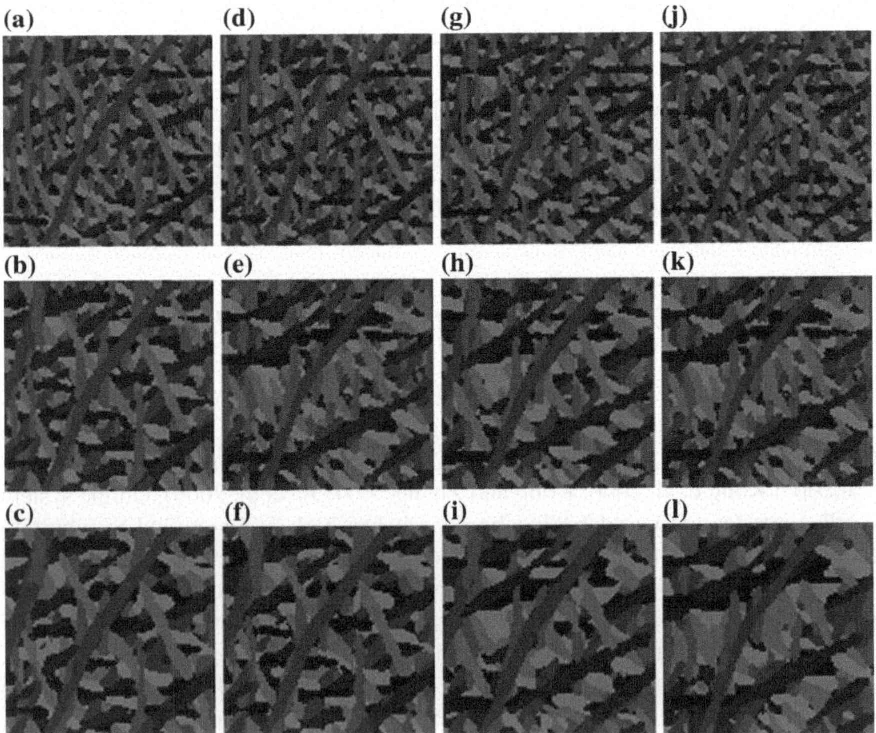

Fig. 6.7 Feature map extracted from Fig. 6.5. Different *gray* values represent different orientation features. The three maps in each column from *left* to *right* are all extracted from the same ROI but using three different parameters. Thus, **a–c** are extracted from blue; **d–f** are extracted from green; and **g–i** are extracted from red. The maps in the fourth column, **j–l**, are extracted from NIR. The same setting is used for each row

Table 6.1 Bit representation of the orientation coding

Orientation value	Bit 0	Bit 1	Bit 2
0	0	0	0
$\pi/6$	0	0	1
$\pi/3$	0	1	1
$\pi/2$	1	1	1
$2\pi/3$	1	1	0
$5\pi/6$	1	0	0

$$D(P,Q) = \frac{\sum_{y=0}^{M}\sum_{x=0}^{N}\sum_{i=1}^{3}\left(P_i^b(x,y)\otimes Q_i^b(x,y)\right)}{3M\times N} \tag{6.2}$$

where P and Q represent two palmprint orientation feature maps, P_i^b and Q_i^b are the ith bit plane of P and Q, respectively. Symbol "$\otimes$" represents bitwise exclusive OR. Obviously, D is between 0 and 1, and for a perfect matching, the distance will be 0.

To further reduce the influence of imperfect ROI extraction, in matching we translate one of the two feature maps vertically and horizontally from −3 to 3. The minimal distance obtained by translated matching is treated as the final distance.

6.3.2 *Inter-spectral Correlation Analysis*

To remove the redundant information of multispectral images, we investigate the correlation between different bands in a quantitative way. By using the Gabor filter in Eq. 6.1, palmprint features are extracted for each spectral band individually, and then the inter-spectral distance is computed by Eq. 6.2 for the same palm. Table 6.2 shows the statistics of inter-spectral distance and Table 6.3 shows that of

Table 6.2 Statistic of inter-spectral distance

Mean/minimal/maximal of distance	Blue	Green	Red	NIR
Blue	0	0.126/0.069/0.302	0.269/0.129/0.404	0.384/0.273/0.470
Green		0	0.252/0.127/0.431	0.371/0.264/0.466
Red			0	0.246/0.148/0.384
NIR				0.384/0.273/0.470

Table 6.3 Statistic of intra-spectral distance

Spectrum	Mean of genuine	Mean of impostor
Blue	0.234	0.459
Green	0.237	0.459
Red	0.221	0.457
NIR	0.237	0.460

intra-spectral distance. From Table 6.2, we can see that as the difference between spectrum increases, the feature difference increases. For example, the average difference between blue and green is 0.1571, while that of blue and NIR is 0.3910. This indicates that different spectra can be used to highlight different textural components of the palm. Meanwhile, compared with the impostor distance, which can be assumed to be independent and close to 0.5 (Daugman 2003), the inter-spectral distance is much smaller, which shows that different spectral bands are correlated rather than independent.

6.3.3 Score-Level Fusion Scheme

Generally, more information is used, better performance could be achieved. However, since there is some overlapping of the discriminating information between different bands, simple sum of the matching scores of all bands may not improve much the final accuracy. Suppose there are *k* kinds of features ($F_i^X,\ i = \{1, 2, \ldots, k\}$). For two samples *X* and *Y*, the distance using simple sum rule is defined as follows:

$$d_{\text{Sum}}(X, Y) = \sum_{i=1}^{k} d(F_i^X, F_i^Y) \tag{6.3}$$

where $d(F_i^X, F_i^Y)$ is the distance for the *i*th feature.

Figure 6.8 shows an example of score-level fusion by summation. There are two kinds of features ($F_i^X,\ i = \{1, 2\}$) for three samples $\{X_1, X_2, Y_1\}$, where X_1 and X_2 belong to the same class and Y_1 belongs to another class. By Eq. 6.3, we can get $d_{\text{Sum}}(X_1, X_2) = 9$ and $d_{\text{Sum}}(X_1, Y_1) = 8$. In fact, the true distances between X_1 and X_2, and X_1 and Y_1 without information overlapping should be 5 and 6, respectively. Because there is an overlapping part between the two features, it will be counted twice by using the sum rule (Eq. 6.3). Sometime, such kind of over-computing may

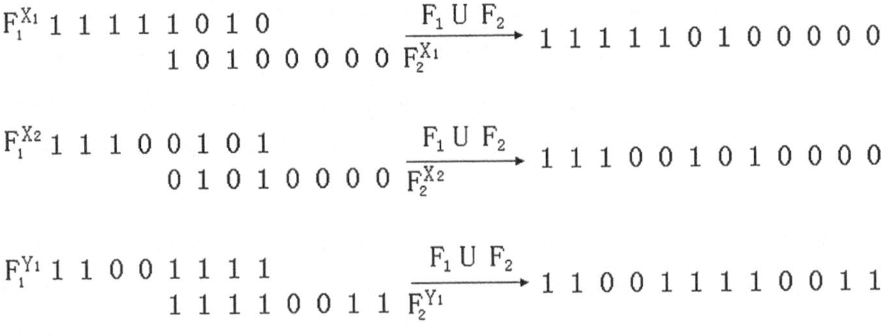

$F_1^{X_1}$ 1 1 1 1 1 0 1 0
1 0 1 0 0 0 0 0 $F_2^{X_1}$ —$F_1 \cup F_2$→ 1 1 1 1 1 0 1 0 0 0 0 0

$F_1^{X_2}$ 1 1 1 0 0 1 0 1
0 1 0 1 0 0 0 0 $F_2^{X_2}$ —$F_1 \cup F_2$→ 1 1 1 0 0 1 0 1 0 0 0 0

$F_1^{Y_1}$ 1 1 0 0 1 1 1 1
1 1 1 1 0 0 1 1 $F_2^{Y_1}$ —$F_1 \cup F_2$→ 1 1 0 0 1 1 1 1 0 0 1 1

Fig. 6.8 An example of sum score fusion

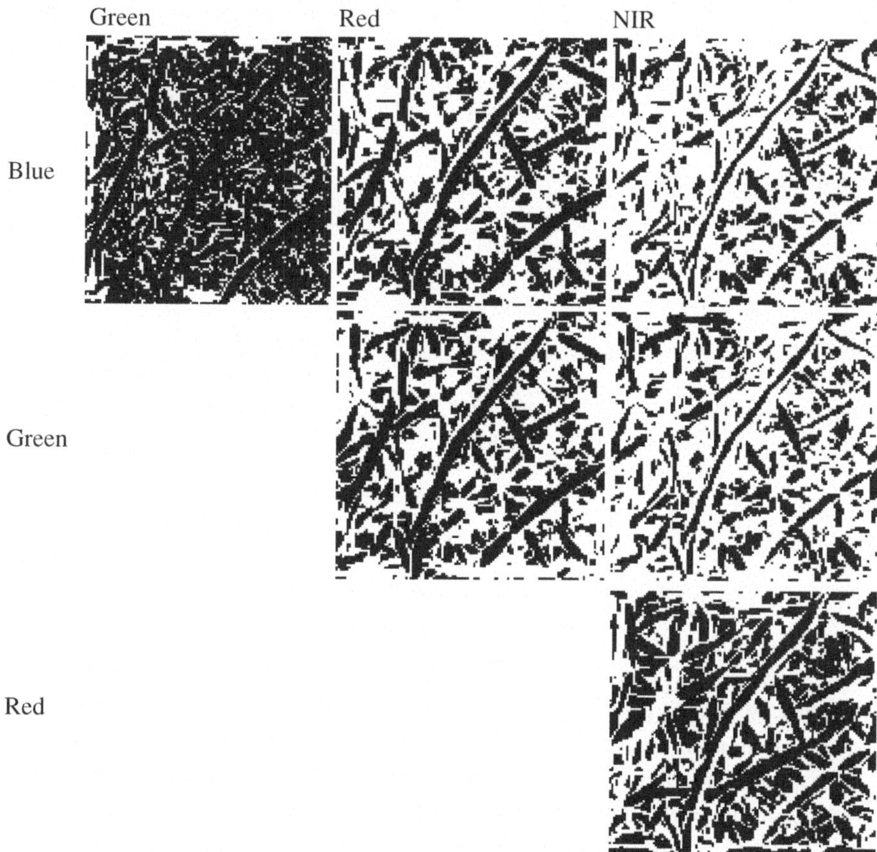

Fig. 6.9 Overlapping features between different spectra. (*Black pixels* represent overlapping features while *white pixels* represent non-overlapping features)

make the simple score-level fusion fail as showed in the above example. For multispectral palmprint images, most of the overlapping features between two spectral bands locate on the principal lines as shown in Fig. 6.9. By using the sum rule (Eq. 6.3), those line features will be over-counted so that it may fail to classify two palms with similar principal lines.

From the above analysis, we can see that if we can find a score-level fusion strategy to reduce the overlapping effect, better verification result can be expected. The "U" operator in the set theory gives us a good hint, which is defined as follows:

$$X \cup Y = X + Y - X \cap Y \tag{6.4}$$

Based on Eq. 6.4, we define a score-level fusion rule which tends to minimize the overlapping effect on the fused score:

Table 6.4 The statistical percentage (%) of overlapping features between and among different spectral bands on half of the database

Spectra	Mean percentage	Standard percentage
Blue and green	72.6535	3.9052
Blue and red	47.0632	4.9618
Blue and NIR	33.6842	5.0552
Green and red	47.3961	4.3286
Green and NIR	34.3067	4.6105
Red and NIR	55.2617	5.5367
Blue, green and red	38.8421	4.6939
Blue, green and NIR	26.9385	4.3295
Blue, red and NIR	26.3367	4.3802
Green, red and NIR	26.8390	4.2041
Blue, green, red and NIR	21.9879	3.9503

$$\begin{aligned} d_{F_1 \cup F_2}(X,Y) &= d(F_1) + d(F_2) - d(F_1 \cap F_2) \\ &= d(F_1^X, F_1^Y) + d(F_2^X, F_2^Y) - \frac{(d(F_1^X, F_1^Y) + d(F_2^X, F_2^Y))}{2} * P_{OP}(F_1, F_2) \end{aligned} \tag{6.5}$$

where $P_{OP}(F_1, F_2)$ is the overlapping percentage between two feature maps. Here, two assumptions are made. First, we assume that the overlapping percentage of two feature maps is nearly the same for different palms. There are two reasons for us to make this assumption. One is that the difference of overlapping percentage between different palms is relatively small, as can be seen in Table 6.4. The other one is that although $P_{OP}(F_1, F_2)$ can be computed for any given two feature maps, it will spend some computational cost and hence may be a burden in time-demanding applications. Thus, to improve the processing speed, we fix $P_{OP}(F_1, F_2)$ as the average value computed from half of the database. The second assumption we made is that the overlapping is uniformly distributed across the feature map. Thus, we can use $\frac{(d(F_1^X, F_1^Y) + d(F_2^X, F_2^Y))}{2} * P_{OP}(F_1, F_2)$ as an approximation distance in the overlapping part.

By using Eq. 6.5, the distances between X_1 and X_2, and X_1 and Y become 6.75 and 6, respectively. It is much closer to the true distance than using Eq. 6.3. Similarly, we could extend the fusion scheme to fuse more bands, e.g., three spectral bands as in Eq. 6.6:

$$
\begin{aligned}
d_{F_1 \cup F_2 \cup F_3}(X,Y) = {} & d(F_1^X, F_1^Y) + d(F_2^X, F_2^Y) + d(F_3^X, F_3^Y) \\
& - \frac{(d(F_1^X, F_1^Y) + d(F_2^X, F_2^Y))}{2} * P_{\mathrm{OP}}(F_1, F_2) \\
& - \frac{(d(F_1^X, F_1^Y) + d(F_3^X, F_3^Y))}{2} * P_{\mathrm{OP}}(F_1, F_3) \\
& - \frac{(d(F_2^X, F_2^Y) + d(F_3^X, F_3^Y))}{2} * P_{\mathrm{OP}}(F_2, F_3) \\
& + \frac{(d(F_1^X, F_1^Y) + d(F_2^X, F_2^Y) + d(F_3^X, F_3^Y))}{3} * P_{\mathrm{OP}}(F_1, F_2, F_3)
\end{aligned}
\tag{6.6}
$$

Because different bands highlight different features of the palm, these features may provide different discriminate capabilities. It is intuitive to use weighted sum:

$$
d_{\text{Weight Sum}} = \sum_{i=1}^{n} w_i d_i \tag{6.7}
$$

where w_i is the weight on d_i, the distance in the ith band, and n is the number of total bands. Equation 6.3 can be regarded as a special case of Eq. 6.7 when the weight is 1 for each spectrum.

Using the reciprocal of equal error rate [EER, a point when false accept rate (FAR) is equal to false reject rate (FRR)] as weight is widely used in Biometric Systems (Snelick et al. 2005). If we take $d_i' = w_i d_i$ as the normalized distance for band i, we can extend our proposed score-level fusion scheme to weighted sum: multiply original distance with the weight for normalization, and then substitute the new distance into Eq. 6.5 or Eq. 6.6.

6.4 Experimental Results

In this section, we report the experimental results on a large database. Section 6.4.1 introduces the establishment of our multispectral palmprint database. Section 6.4.2 reports the verification results on each band. Section 6.4.3 reports the results using proposed fusion scheme. Section 6.4.4 shows some simple anti-spoofing tests. Finally, Sect. 6.4.5 discusses the speed.

6.4.1 Multispectral Palmprint Database

We collected multispectral palmprint images from 250 individuals using the developed data acquisition device. The subjects were mainly volunteers from our institute. In the database, 195 people are male and the age distribution is from 20 to 60 years old. We collected the multispectral palmprint images on two separate sessions. The average time interval between the two occasions is 9 days. On each session, the subject was asked to provide six samples of each of his/her left and right palms. So our database contains 6000 images for each band from 500 different palms. For each shot, the device collected four images from the four bands (red, green, blue, and NIR) in less than one second. In palmprint acquisition, the users are asked to keep their palms stable on the device. The resolution of the images is 352 by 288 (<100 dpi).

6.4.2 Palmprint Verification on Each Band

As shown in Fig. 6.5, different bands can have different features of palm, providing different discriminate information for personal authentication. For different bands, different parameters should be used for better recognition results. This section is to investigate which spectral band works better for palmprint verification. Specifically, we try to find which band is preferred for palmprint verification via an experimental point of view.

In the following experiment, we set up a subset from the whole database by using the first session images only, totally 3000 images for each band. The subset is used for parameter selection in feature extraction. To obtain the verification accuracy, each palmprint image is matched with all the other palmprint images of the same band in the training database. A match is counted as a genuine matching if the two palmprint images are from the same palm; otherwise, the match is counted as impostor matching. The total number of matches is $3000 \times 2999/2 = 4{,}498{,}500$, and among them, there are 7500 ($6 \times 500 \times 5/2$) genuine matching and others are impostor matching. The EER is used to evaluate the performance.

As stated in Sect. 6.3, there are two variables for the Gabor filter: κ and ω. Different parameters may get different feature maps as shown in Fig. 6.5. It is impossible to search exhaustively in the parameter space to find the optimal parameters for each band. We evaluate on 20 and 15 different values for κ and ω; thus, the total number of combination trial is 300. Here, the Gabor filter size is fixed as 35×35.

Some statistics of the EER are listed in Table 6.5. Due to the limit of space, we do not show all the experiment results. Because of the different spectral characteristics of different bands, the optimal (corresponding to minimal EER) parameters for different bands are different. Some findings could be found from Table 6.5. Firstly, red and NIR bands have better EER than blue and green bands. This is mainly because red and NIR could not only capture most of the palm line information, but also capture some palm vein structures. This additional palm vein information helps classify those palms with similar palm lines. Figure 6.10 shows an example, the two multispectral images of different palms have small distance under blue [(a) and (e)] or green [(b) and (f)] bands, which may lead to a false

Table 6.5 Statistics of EER under different parameters for different bands

EER (%)	Blue	Green	Red	NIR
Mean	0.0712	0.0641	0.0257	0.0430
Std	0.0142	0.0170	0.0217	0.0340
Median	0.0669	0.0666	0.0153	0.0378
Minimal	0.0400	0.0293	0.0015	0.0114
Maximal	0.1332	0.1607	0.1601	0.2665

match; however, their distance under red [(c) and (g)] or NIR [(d) and (h)] bands is large, which will lead to a correct match.

Secondly, the EER of NIR is higher than that of red. There are mainly two reasons. First, the palm lines in NIR band are not as strong as those in the red band because NIR light can penetrate deeper the palm skin than red light, which attenuates the reflectance. Second, some people, especially females, have very weak vein structures under NIR light because their skin is a little thicker (Lee and Hwang 2002). Figure 6.11 shows an example. Figures 6.11b, d are the NIR images of a palm. The line structure in the two images is very weak compared with Figs. 6.11a, c, which are the red images of the palm. Meanwhile, the vein structure in Figs. 6.11b, d is not strong enough. Thus, the palm will be falsely recognized in the NIR band, while it can be recognized in the red band. Thirdly, the performance of blue and green bands is very similar. As can be seen in Figs. 6.5 and 6.10, the palm lines in blue and green bands look very similar. Last but not the least, the accuracy on short-wavelength bands (blue and green) is much more robust to parameter selection than that on long-wavelength bands (red and NIR). One possible reason is

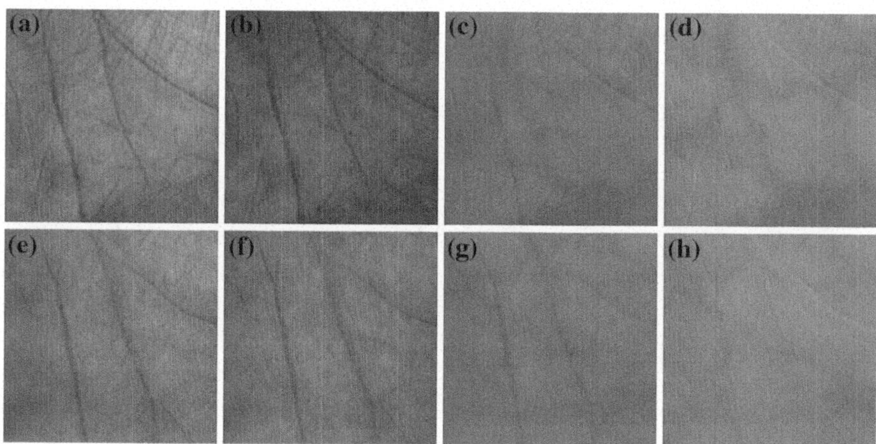

Fig. 6.10 Multispectral images of two different palms which are wrongly recognized under blue or green spectrum, but can be correctly recognized under red or NIR spectrum. From **a–b** or **e–h**, the four images correspond to blue, green, red, and NIR respectively

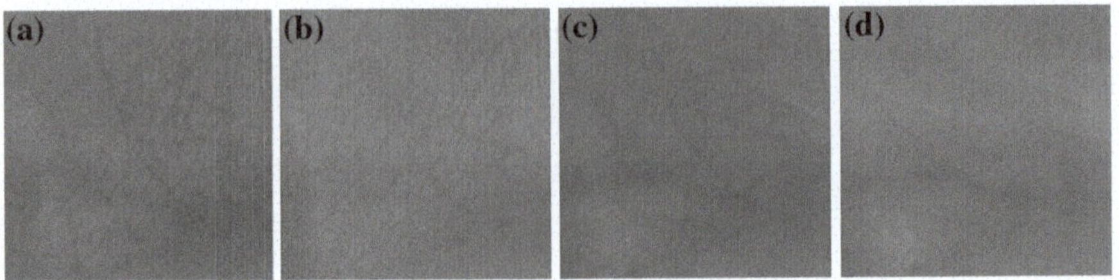

Fig. 6.11 An example of palm which is falsely recognized in NIR band but can be correctly recognized in red band. **a** and **c** are collected under red light while **b** and **d** are collected under NIR light

that short-wavelength light scatters more than long-wavelength light (Bohren and Huffman 1983), so palm lines are much thicker and clearer under short spectra.

We perform the palmprint verification on the whole database with the optimal parameters for each band. The same matching strategy is used as before. The numbers of genuine matching and impostor matching are 33,000 and 17,964,000, respectively. The receiver operating characteristic (ROC) curves for different spectral bands are shown in Fig. 6.12. The EER values for blue, green, red and NIR are 0.0520, 0.0575, 0.0212, and 0.0398 %, respectively. The accuracy on each single band is comparable to those of state of the art (EER: 0.024 %) (Zuo et al. 2008) on the public palmprint database (PolyU Palmprint Database) collected under white illumination.

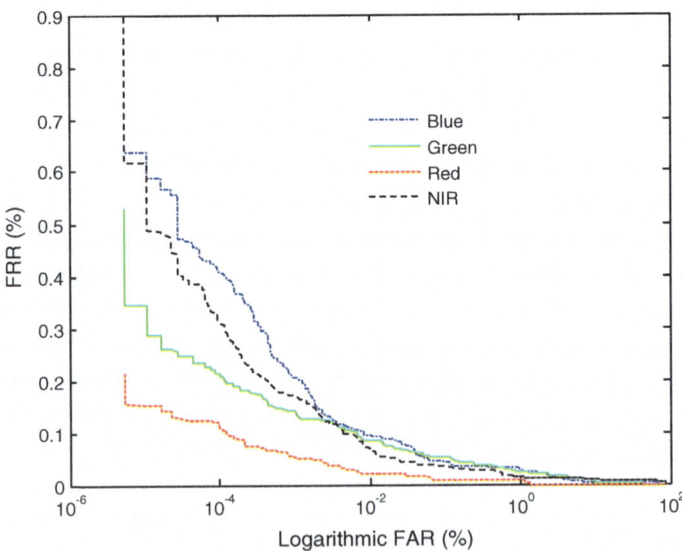

Fig. 6.12 Verification performance using optimal parameters for each spectrum

6.4.3 Palmprint Verification by Fusion

The four bands can be fused to further improve the palmprint verification accuracy. Table 6.6 lists the accuracy results by four different fusion schemes, original weighted sum (w = 1), proposed weighted sum (w = 1), original weighted sum (w = 1/EER), and proposed weighted sum (w = 1/EER). Some findings could be obtained. Firstly, all fusion schemes can result in smaller EER than a single band except the fusion of blue and green (this is because the feature overlapping between them is very high), which validate the effectiveness of multispectral palmprint authentication. Secondly, using the reciprocal of EER as weight usually leads to better results than the equal weight scheme. Thirdly, the proposed fusion scheme, which could reduce the feature overlapping effect, achieves better results than the original weighted sum method. It can be verified that Eq. 6.5 can be rewritten as Eq. 6.7, and it is actually a weighted $((1 - P_{\mathrm{OP}}(F_1, F_2)))$ distance of Eq. 6.3.

$$\begin{aligned} d_{F_1 \cup F_2}(X, Y) &= d(F_1) + d(F_2) - d(F_1 \cap F_2) \\ &= d(F_1^X, F_1^Y)(1 - P_{\mathrm{OP}}(F_1, F_2)) + d(F_2^X, F_2^Y)(1 - P_{\mathrm{OP}}(F_1, F_2)) \\ &= (d(F_1^X, F_1^Y) + d(F_2^X, F_2^Y))(1 - P_{\mathrm{OP}}(F_1, F_2)) \end{aligned} \tag{6.8}$$

Table 6.6 Accuracy measurement comparison by different fusion schemes

EER (%)	Original weighted sum (w = 1)	Proposed weighted sum (w = 1)	Original weighted sum (w = 1/EER)	Proposed weighted sum (w = 1/EER)
Blue, green	0.0425	0.0425	0.0397	0.0397
Blue, red	0.0154	0.0154	0.0121	0.0121
Blue, NIR	0.0212	0.0212	0.0212	0.0212
Green, red	0.0212	0.0212	0.0182	0.0182
Green, NIR	0.0242	0.0242	0.0181	0.0181
Red, NIR	0.0152	0.0152	0.0152	0.0152
Blue, green, red	0.0243	0.0212	0.0152	0.0151
Blue, green, NIR	0.0212	0.0214	0.0212	0.0212
Blue, red, NIR	0.0121	0.0121	0.0121	0.0121
Green, red, NIR	0.0153	0.0156	0.0152	0.0150
Blue, green, red, NIR	0.0152	0.0151	0.0121	0.0121

The best result is obtained by fusing red and blue bands, which leads an EER as low as 0.0121 %. To the best of our knowledge, our multispectral palmprint database (250 subjects) is the largest database so far. The numbers of subjects in the databases of Hao et al. (2007), Rowe et al. (2007), Likforman-Sulem et al. (2007), Wang et al. (2008), and Hao et al. (2008) are 7, 50, 100, 120, and 165, respectively. Among them, only the size of the database by Hao et al. (2008) is close with ours. However, the best EER of their work is 0.50 %, which is much worse than ours (0.0121 %).

Although our fusion scheme verifies that the accuracy could be improved by fusing the features across spectral bands, sometimes the fusion of three or four bands is not better than the fusion of two bands. Certainly, it is possible that a better fusion scheme could be developed to more efficiently fuse the different features in different bands. How to further improve the fusion result will be explored in our future work.

6.4.4 Anti-spoofing Test

A good biometric system should be robust to spoof attacks. To test the anti-spoofing ability, a simple test is implemented. A palmprint image in blue band is printed on a paper, and then, this paper is used as a fake palm to attack the system. For comparison, we apply this test to both the single-spectral (i.e., the blue channel) system and multispectral system. As shown in Fig. 6.13, the faked palmprint is easy to pass the single-spectral system, the distance (0.3767) between (a) and (e) is smaller than the mean impostor distance (0.4621, refer to Table 6.2 please) and close to the verification threshold in our system. But it is hard to pass the multispectral system because the distances in other bands are big. For example, the distance (0.4426) between (d) and (h) is close to the mean impostor distance (0.4627) and far away from the verification threshold. Moreover, because the reflectance of the fake material (paper in this example) is different from that of the skin, the distance between blue and NIR bands of the faked palm is very small compared with that of the true palm (refer to Table 6.2 please). Thus, this feature can be used for liveness detection to improve the robustness of our system.

6.4.5 Speed

The system is implemented using Visual C++6.0 on a PC with Windows XP, T6400 CPU (2.13 GHz) and 2 GB Ram. The execution time for each step is listed in Table 6.7. The total execution time of verification is less than 1.3 s, which is fast enough for real-time application. As the speed of matching is fast, it can be easily extended to identification system. For example, for 1–1000 identification, the total

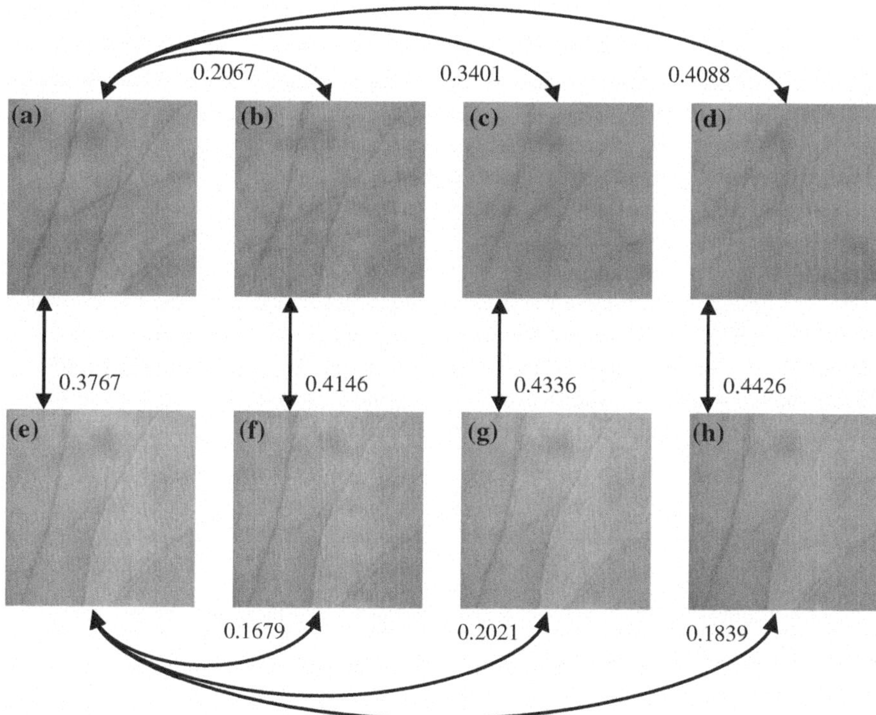

Fig. 6.13 An example of anti-spoof test. **a–d** are images of a true palm taken under blue, green, red, and NIR. **e–h** are images of a fake palm taken under blue, green, red, and NIR. The distance between two images is showed on or near a *double-arrow curve*

Table 6.7 Execution time

	Average time (ms)
Image acquisition	<1000
ROI extraction	138
Feature extraction	36 × 4
Feature matching	0.06 × 4

time cost is only 1.5 s. By optimizing the program code, it is possible to further reduce the computational time.

6.5 Summary

In this chapter, we proposed an online multispectral palmprint system for real-time biometric authentication. Using palm line orientation code as features, we first evaluated palmprint verification using single spectrum. It was verified that red and

NIR bands have better results than blue and green bands because they could not only capture palm line features but also palm vein features. Since the palm lines acquired from NIR are not as clear as those from red and the palm vein structure obtained from NIR solely is not discriminate enough, NIR is a little inferior to red. The blue and green bands have very similar results and they are more robust to parameter selection than long-wavelength bands (red and NIR bands).

Since different bands highlight different texture information, the fusion of them could reduce EER significantly. It was found that the fusion of red and blue achieves the best result. A new score-level fusion scheme originated from the set theory was proposed to reduce the overlapping effect between bands. Experimental results show the effectiveness of our scheme. It was also found that due to the much redundant information across some bands, fusion of three or four bands may not achieve better result than the fusion of two bands. How to better fuse the multispectral information will be investigated in the future.

References

Bohren CF, Huffman D (1983) Absorption and scattering of light by small particles. Wiley, New York

Connie T, Andrew T, Goh K (2005) An automated palmprint recognition system. Image Vis Comput 23(5):501–505

Daugman J (2003) The importance of being random: statistical principles of iris recognition. Pattern Recogn 36(2):279–291

Gawkrodger DJ (2002) Dermatology: an illustrated colour text, 3rd ed. Elsevier, Amsterdam

Han C, Cheng H, Lin C, Fan K (2003) Personal authentication using palm-print features. Pattern Recogn 36(2):371–381

Hao Y, Sun Z, Tan T (2007) Comparative studies on multispectral palm image fusion for biometrics. In: Asian conference on computer vision, pp 12–21

Hao Y, Sun Z, Tan T, Ren C (2008) Multispectral palm image fusion for accurate contact-free palmprint recognition. In: International conference on image processing, pp 281–284

Hu D, Feng G, Zhou Z (2007) Two-dimensional locality preserving projections (2DLPP) with its application to palmprint recognition. Pattern Recogn 40(1):339–342

Jain A, Bolle R, Pankanti S (1999) Biometrics: personal identification in network society. Kluwer, Boston

Jia W, Huang D-S, Zhang D (2008) Palmprint verification based on robust line orientation code. Pattern Recogn 41(5):1504–1513

Kong A, Zhang D, Kamel M (2006) Palmprint identification using feature-level fusion. Pattern Recogn 39(3):478–487

Kong A, Zhang D (2004) Competitive coding scheme for palmprint verification. Int Conf Pattern Recogn 1:520–523

Lee Y, Hwang K (2002) Skin thickness of Korean adults. Surg Radiol Anat 24(3–4):183–189

Likforman-Sulem L, Salicetti S, Dittmann J, Ortega-Garcia J, Pavesic N, Gluhchev G, Ribaric S, Sankur B (2007) Final report on the jointly executed research carried out on signature, hand and other modalities. http://www.cilab.upf.edu/biosecure1/public_docs_deli/BioSecure_Deliverable_D07-4-4_b2.pdf.pdf

Park JH, Kang MG (2007) Multispectral iris authentication system against counterfeit attack using gradient-based image fusion. Opt Eng 46(11):117003-1–117003-14

Ross AA, Nadakumar K, Jain AK (2006) Handbook of multibiometrics. Springer, Berlin

Rowe RK, Nixon KA, Corcoran SP (2005) Multispectral fingerprint biometrics. In: Proceedings of information assurance workshop, pp 14–20

Rowe RK, Uludag U, Demirkus M, Parthasaradhi S, Jain AK (2007) A multispectral whole-hand biometric authentication system. In: Proceedings of biometric symposium. biometric consortium conference. Baltimore, September

Schukers SAC (2002) Spoofing and anti-spoofing measures. Inf Secur Tech Rep 7(4):56–62

Singh R, Vatsa M, Noore A (2008) Hierarchical fusion of multispectral face images for improved recognition performance. Inf Fusion 9(2):200–210

Snelick R, Uludag U, Mink A, Indovina M, Jain A (2005) Large-scale evaluation of multimodal biometric authentication using state-of-the-art systems. IEEE Trans Pattern Anal Mach Intell 27(3):450–455

Wang J-G, Yau W-Y, Suwandy A, Sung E (2008) Person recognition by fusing palmprint and palm vein images based on "Laplacianpalm" representation. Pattern Recogn 41(5):1514–1527

Wu X, Zhang D, Wang K (2003) Fisherpalm based palmprint recognition. Pattern Recogn Lett 24 (15):2829–2838

Wu X, Zhang D, Wang K (2006) Palm line extraction and matching for personal authentication. IEEE Trans Syst Man Cybern Part A 36(5):978–987

Zhang D (2000) Automated biometrics—technologies and systems. Kluwer, Boston

Zhang D, Kong W, You J, Wong M (2003) Online palmprint identification. IEEE Trans Pattern Anal Mach Intell 25(9):1041–1050

Zharov VP, Ferguson S, Eidt JF, Howard PC, Fink LM, Waner M (2004) Infrared imaging of subcutaneous veins. Lasers Surg Med 34(1):56–61

Zuo W, Yue F, Wang K, Zhang D (2008) Multiscale competitive code for efficient palmprint recognition. In: International conference on pattern recognition

Chapter 7
Empirical Study of Light Source Selection for Palmprint Recognition

Abstract Most of the current palmprint recognition systems use an active light to acquire images, and the light source is a key component in the system. Although white light is the most widely used light source, little work has been done on investigating whether it is the best illumination for palmprint recognition. This study analyzes the palmprint recognition performance under seven different illuminations, including the white light. The experimental results on a large database show that white light is not the optimal illumination, while yellow or magenta light could achieve higher palmprint recognition accuracy than the white light.

Keywords Biometrics · Palmprint recognition · Illumination · Orientation code

7.1 Introduction

Automatic authentication using biometric characteristics, as a replacement or complement to traditional personal authentication, is becoming more and more popular in the current e-world. Biometrics is the study of methods for uniquely recognizing humans based on one or more intrinsic physical or behavioral traits, such as face, iris, fingerprint, signature, gait, and finger-knuckle-print (Jain et al. 1999; Zhang et al. 2010a, b). As an important member of the biometric characteristics, palmprint has merits such as robustness, user-friendliness, high accuracy, and cost-effectiveness. Because of these good properties, palmprint recognition has been receiving a lot of research attention and many palmprint systems have been proposed (Connie et al. 2005; Duta et al. 2002; Han 2004; Han et al. 2007; Kumar et al. 2003; Lin et al. 2005; Michael et al. 2008; Ribaric and Fratric 2005; Wang et al. 2008; Wu et al. 2008; Zhang et al. 2003; Zhang and Shu 1999).

D. Zhang et al., *Multispectral Biometrics*,
DOI 10.1007/978-3-319-22485-5_7

In the early stage, most works focus on off-line palmprint images (Duta et al. 2002; Zhang and Shu 1999).With the development of digital image acquisition devices, many online palmprint systems have been proposed (Connie et al. 2005; Han 2004; Han et al. 2007; Kumar et al. 2003; Lin et al. 2005; Michael et al. 2008; Ribaric and Fratric 2005; Wang et al. 2008; Wu et al. 2008; Zhang et al. 2003). Based on the sensors used, online palmprint image acquisition systems can be grouped into four types (Kong et al. 2009): digital scanners (Connie et al. 2005; Lin et al. 2005; Ribaric and Fratric 2005), video cameras (Han et al. 2007; Michael et al. 2008), CCD (charge-coupled device)-based palmprint scanner (Han 2004; Wang et al. 2008; Zhang et al. 2003), and digital cameras (Kumar et al. 2003; Wu et al. 2008). On the other hand, according to imaging conditions, these systems could be classified into three classes: digital scanners (Connie et al. 2005; Lin et al. 2005; Ribaric and Fratric 2005), camera with passive illumination (Han et al. 2007; Kumar et al. 2003; Wang et al. 2008), and camera with active illumination (Han 2004; Michael et al. 2008; Wu et al. 2008; Zhang et al. 2003).

Desktop scanner could provide high-quality palmprint images (Connie et al. 2005; Lin et al. 2005; Ribaric and Fratric 2005) under different resolutions. However, it suffers from the slow scanning speed (Kong et al. 2009), and it requires the full touch of whole hand, which may bring sanitary issues during data collection. Using camera with uncontrolled ambient lighting (Han et al. 2007; Kumar et al. 2003; Wang et al. 2008) does not have the above problems. However, the image quality may not be very good as the illumination lighting may change much so that the recognition accuracy can be reduced. Because camera mounted with active light could collect image data quickly with good image quality and it does not require the full touch with the device, this kind of systems has been widely adopted (Han 2004; Michael et al. 2008; Wu et al. 2008; Zhang et al. 2003). In these systems, the light source is a key component and there are some principles on the setting of lighting scheme (Wong et al. 2005). In spite of the fact that all these studies (Han 2004; Michael et al. 2008; Wu et al. 2008; Zhang et al. 2003) use white light source for palmprint imaging, little work has been done to systematically validate whether white light is the optimal light. Our previous work (Guo et al. 2009) showed that white light may not be the optimal one for palmprint recognition, but the finding may be biased as the study is limited by using only one recognition method. Furthermore, the underlying principle is not explored. To this end, this study discusses the light selection for palmprint recognition through extensive experiments on a large multispectral palmprint database we established (Han et al. 2008).

In general, there are three kinds of popular approaches to palmprint recognition: structural methods (Lin et al. 2005; Liu 2007), statistical methods (Zhang and Zhang 2004; Connie et al. 2005; Han 2004; Ribaric and Fratric 2005; Wang et al. 2008), and texture-based coding (Han et al. 2007; Kong and Zhang 2004; Michael et al. 2008; Zhang et al. 2003) methods. Different approaches focus on different kinds of features. For example, structural methods pay attention to principal lines and wrinkles (Zhang et al. 2003), statistical methods use holistic expression for palmprint recognition, and texture coding schemes assign a feature code for each pixel in the palmprint. Different illuminations may enhance different palmprint features as the reflectance and absorbance of human skin relate with spectrum, for example, short-wavelength light could enhance the line feature of palms, while long-wavelength light could acquire some subcutaneous vein structure (Angelopoulou 2001). Thus, to get unbiased result for different illuminations, this study employs three different kinds of feature extraction methods, i.e., Competitive Coding (Kong and Zhang 2004) (texture-based coding), wide line detection (Liu 2007) (structural method), and $(2D)^2$PCA (Zhang and Zhou 2005) (statistical method), to empirically study the light source selection for palmprint recognition.

The rest of this chapter is organized as follows. Section 7.2 describes our data collection. Section 7.3 briefly introduces the three feature extraction algorithms. Section 7.4 presents the light source analysis results, and Sect. 7.5 concludes the chapter.

7.2 Multispectral Palmprint Data Collection

It is known that red, green, and blue are the three primary colors, and the combination of them could result in many different colors in the visible spectrum. We established a multispectral palmprint data collection device (introduced in Chap. 6) which includes the three primary color illumination sources (LED light sources). By using this device, we can simulate different illumination conditions. For example, when the red and green LEDs are switched on simultaneously, the yellow-like light could be generated. Totally, our device could collect palmprint images under seven different color illuminations: red, green, blue, cyan, yellow, magenta, and white. Figure 7.1 shows examples of the collected images under different illuminations.

Fig. 7.1 Sample images of three different palms with different illuminations. Images of each column come from the same palm and are collected by red, green, blue, yellow, magenta, cyan, and white illuminations

7.3 Feature Extraction Methods

7.3.1 Wide Line Detection

A palmprint image has mainly three kinds of features: principal lines (usually three dominant lines on the palm), wrinkles (weaker and more irregular lines), and crease (the ridge and valley structures like in fingerprint) (Zhang et al. 2003). These principal lines and wrinkles could be extracted by a wide line detector directly (Liu 2007). The thickness of the line is defined as:

$$\begin{aligned} L(x_0,y_0) &= \begin{cases} g-m(x_0,y_0) & \text{if } m(x_0,y_0)<g \\ 0 & \text{otherwise} \end{cases} \\ m(x_0,y_0) &= \sum_{x_0-r\le x\le x_0+r,\, y_0-r\le y\le y_0+r} c(x,y,x_0,y_0),\ c(x,y,x_0,y_0) \\ &= \omega_0^* \begin{cases} 0 & \text{if } I(x,y)>I(x_0,y_0) \\ 1 & \text{otherwise} \end{cases} \\ \omega_0 &= \frac{\omega}{\sum_{x_0-r\le x\le x_0+r,\, y_0-r\le y\le y_0+r} \omega(x,y,x_0,y_0,r)} \quad \text{and} \quad \omega(x,y,x_0,y_0,r) \\ &= \begin{cases} 1 & \text{if } (x-x_0)^2+(y-y_0)^2\le r^2 \\ 0 & \text{otherwise} \end{cases} \end{aligned} \tag{7.1}$$

where g is the geometric threshold, r is radius of the circular mask, and m is the weighed mask having similar brightness. ω is a circular constant weighing mask, and ω_0 is the normalization of the circular mask.

To remove noise, a Gaussian smoothing process is employed as a post-processing step. Then, the response is binarized after thresholding:

$$\tilde{L}(x,y) = L(x,y)^* g_\sigma(x,y) \quad \text{where} \quad g_\sigma(x,y) = \frac{1}{2\pi\sigma^2}\exp\left(-\frac{x^2+y^2}{2\sigma^2}\right) \tag{7.2}$$

$$B(x,y)=\begin{cases} 1 & \text{if } \tilde{L}(x,y)>t \\ 0 & \text{otherwise} \end{cases} \tag{7.3}$$

where σ is the scale of the Gaussian filter, and t is a threshold. After binarization, the similarity between two palmprints is defined as the proportion of matched bits to the total bits of the two binary palmprint maps (Liu 2007). As shown in Fig. 7.2, different parameters will generate different feature maps. In the experiments, we will compute the recognition accuracy on a group of parameters and select the statistical value of each illumination for final comparison.

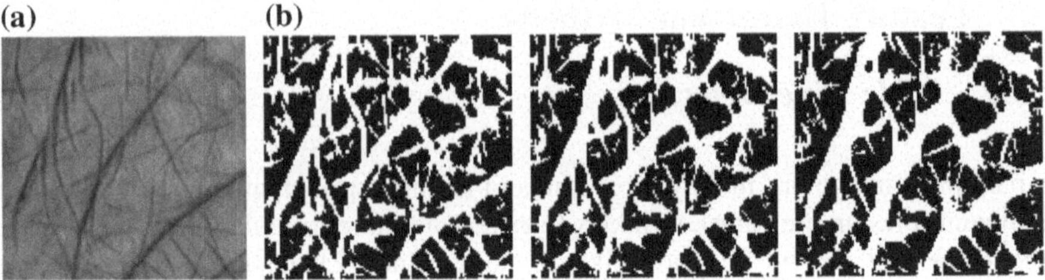

Fig. 7.2 A palmprint ROI (region of interest) sample and its extracted features by wide line detection. **a** A ROI sample. **b** Extracted features by different parameters

7.3.2 Competitive Coding

The orientation of palm lines is stable and can serve as distinctive features for personal identification (Kong and Zhang 2004). As described in Chap. 6, competitive coding scheme is also studied in this Chapter.

7.3.3 (2D)²PCA

Principal component analysis (PCA) is a widely used statistical analysis method, and (2D)²PCA (Zhang and Zhou 2005) is an extension of it, which can alleviate much the small sample size problem and better preserve the image local structural information. Suppose we have M subjects and each subject has S sessions in the training dataset, i.e., S multispectral palmprint cube was acquired at different times for each subject. Then, we denote by X_{ms}^{b} (the original image matrix) the bth band image for the mth individual in the sth session. The covariance matrices along the row and column directions are computed as follows:

$$G_1^b = \frac{1}{MS}\sum_{s=1}^{S}\sum_{m=1}^{M}(X_{ms}^b - \overline{X^b})^T(X_{ms}^b - \overline{X^b}),$$
$$G_2^b = \frac{1}{MS}\sum_{s=1}^{S}\sum_{m=1}^{M}(X_{ms}^b - \overline{X^b})(X_{ms}^b - \overline{X^b})^T \tag{7.4}$$

where $\overline{X^b} = \frac{1}{MS}\sum_{s=1}^{S}\sum_{m=1}^{M}X_{ms}^b$.

The project matrix $V_1^b = [v_{11}^b, v_{12}^b, \ldots, v_{1k_1^b}^b]$ is composed of the orthogonal eigenvectors of G_1^b corresponding to the k_1^b largest eigenvalues, and the projection matrix $V_2^b = [v_{21}^b, v_{22}^b, \ldots, v_{2k_2^b}^b]$ consists of the orthogonal eigenvectors of G_2^b corresponding to the largest k_2^b eigenvalues. k_1^b and k_2^b can be determined by setting a threshold to the cumulant eigenvalues:

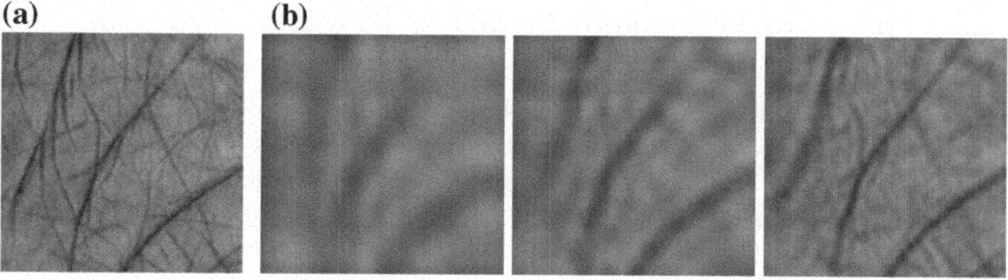

Fig. 7.3 A ROI sample with its reconstructed images by (2D)[2]PCA. **a** A ROI sample. **b** Reconstructed images by different cumulant values

$$\sum_{j_c=1}^{k_1^b} \lambda_{1j_c}^b \Big/ \sum_{j_c=1}^{I_c} \lambda_{1j_c}^b \geq C_u, \sum_{j_r=1}^{k_2^b} \lambda_{2j_c}^b \Big/ \sum_{j_r=1}^{I_r} \lambda_{2j_c}^b \geq C_u \tag{7.5}$$

where $\lambda_{11}^b, \lambda_{12}^b, \ldots, \lambda_{1I_c}^b$ are the first I_c biggest eigenvalues of G_1^b, $\lambda_{21}^b, \lambda_{22}^b, \ldots, \lambda_{2I_r}^b$ are the first I_r biggest eigenvalues of G_2^b, and C_u is a preset threshold. For each given band bth, the test image T^b is projected to $\widetilde{T^b}$ by V_1^b and V_2^b ($V_2^{bT} \times T^b \times V_1^b$), and then, Euclidean distance is used to measure the dissimilarity (Zhang and Zhou 2005). Figure 7.3 shows the reconstructed image ($V_2^b \times \widetilde{T^b} \times V_1^{bT}$) by different C_u.

7.4 Analyses of Light Source Selection

7.4.1 Database Description

We collected multispectral palmprint images from 250 subjects using the developed data acquisition device. The subjects were mainly volunteers from our institutes. In the database, 195 subjects are male and the age distribution is from 20 to 60 years old. We collected the multispectral palmprint images on two separate sessions. The average time interval between the two occasions is 9 days. On each session, the subject was asked to provide 6 samples of each of his/her left and right palms. So our database contains 6000 images for each band from 500 different palms. For each shot, the device collected 7 images from different bands (red, green, blue, cyan, yellow, magenta, and white) in <2 s. In palmprint acquisition, the users are asked to keep their palms stable on the device. The resolution of the images is 352 × 288 (<100 DPI).

After obtaining the multispectral cube, a local coordinate of the palmprint image is established (Zhang et al. 2003) from the blue band, and then, a ROI is cropped from each band based on the local coordinate. For the convenience of analysis, we normalized these ROIs to a size of 128 × 128. To remove the global intensity and contrast effect (Zuo et al. 2006), all images are normalized to have a mean of 128 and standard deviation of 20.

7.4.2 Palmprint Verification Results by Wide Line Detection

To compute the verification accuracy, each palmprint image is matched with all the other palmprint images in the database. A match is counted as a genuine if the two palmprint images are from the same palm; otherwise, it is counted as an impostor. The total number of matches is 17,997,000, and the number of genuine is 33,000. The equal error rate (EER) [the point when false accept rate (FAR) is equal to false reject rate (FRR)] is used to evaluate the accuracy.

As discussed in Sect. 7.3.1, there are four parameters which could influence the feature extraction: r, t, g, and σ. To reduce the possible parameter space, we fixed $g = 0.5$ and $t = 0.1$ (Liu 2007) and selected 9 different values for r ([12, 20]) and 6 different values ([0.75, 2]) for σ, because these values could include optimal parameters (Liu 2007). Thus, the total number of test settings is 54. The EER under different settings with different illuminations is plotted in Fig. 7.4, and the statistical values of EER for each light are listed in Table 7.1.

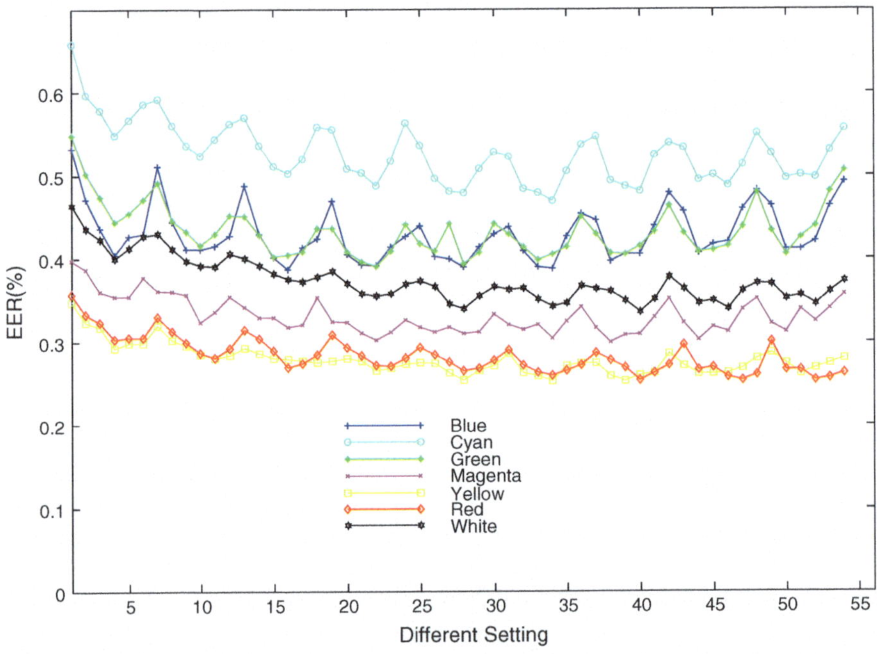

Fig. 7.4 EER values under different settings by wide line detection

Table 7.1 Statistical values of EER for each color by wide line detection

	Blue	Cyan	Green	Magenta	Yellow	Red	White
Mean	0.4306	0.5275	0.4344	0.3318	0.2798	0.2839	0.3738
Standard deviation	0.0328	0.0370	0.0319	0.0221	0.0185	0.0224	0.0278

7.4.3 *Palmprint Verification Results by Competitive Coding*

As discussed in Sect. 7.3.2, parameters ω and δ could influence the extracted features. We selected 20 different values for ω ([4.0, 5.9]) and 15 different values for δ ([1, 1.7]) because these values could cover optimal parameters (Zhang et al. 2010a, b), and we used the same test protocol as discussed in Sect. 7.4.2. Thus, the total number of test settings is 300. The EER under different setting with different illuminations is plotted in Fig. 7.5, and the statistical values of EER for each light are listed in Table 7.2.

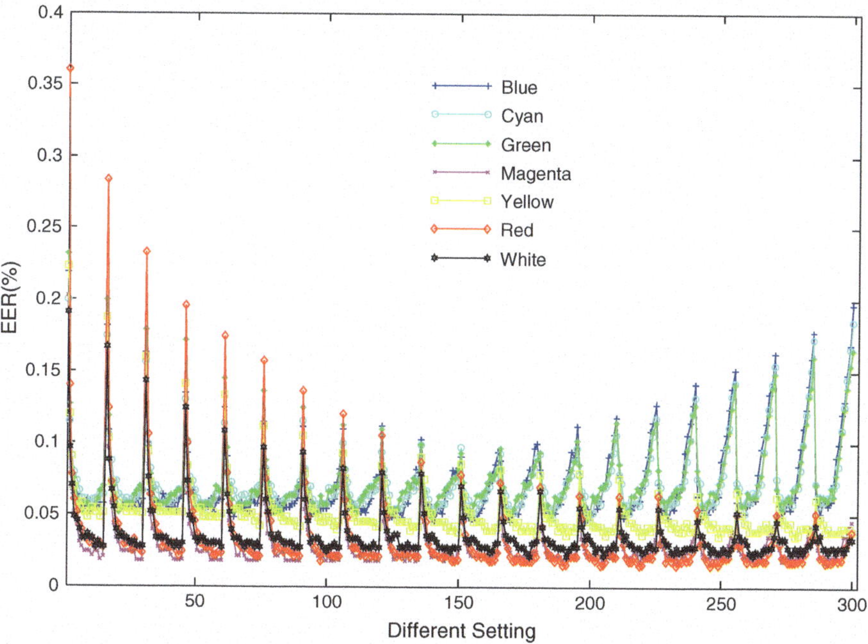

Fig. 7.5 EER values under different settings by competitive coding

Table 7.2 Statistical values of EER for each color by Competitive Coding

	Blue	Cyan	Green	Magenta	Yellow	Red	White
Mean	0.0718	0.0733	0.0742	0.0305	0.0523	0.0351	0.0358
Standard deviation	0.0288	0.0243	0.0262	0.0220	0.0204	0.0358	0.0195

7.4.4 Palmprint Identification Results by $(2D)^2$PCA

In this section, identification instead of verification is implemented. The whole database is partitioned into two parts: training set and test set. The training set is used to estimate the projection matrix and is taken as gallery samples. The test samples are matched with the training samples, and nearest neighborhood classification is employed. The ratio of the number of correct matches to the number of test samples, i.e., the recognition accuracy, is used as the evaluation criteria. To reduce the dependency of experimental results on training sample selection, we designed the experiments as follows. Firstly, the first three samples in the first session are chosen as training set, and the remaining samples are used as test set. Secondly, the first three samples in the second session are chosen as training set, and the remaining samples are used as test set. Finally, the average accuracy is computed.

As shown in Sect. 7.3.3, there is only one parameter, C_u, to control the feature extraction. The accuracies under different settings with different illuminations are plotted in Fig. 7.6, and the highest accuracy for each light is listed in Table 7.3.

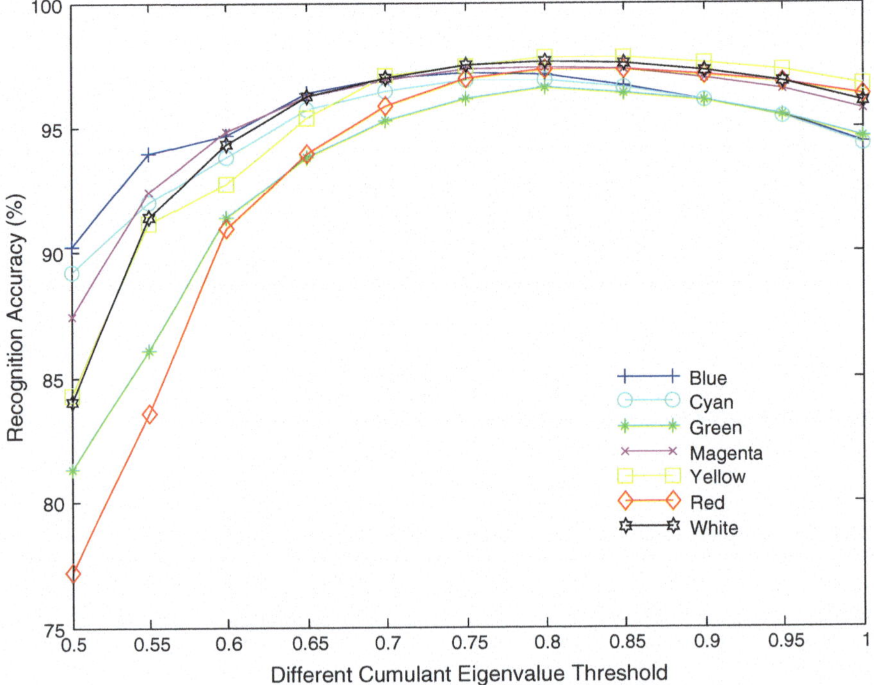

Fig. 7.6 Recognition accuracy under different C_u by $(2D)^2$PCA

Table 7.3 The highest accuracy for each color by $(2D)^2$PCA

Blue	Cyan	Green	Magenta	Yellow	Red	White
97.2000	96.8777	96.6334	97.4333	97.8777	97.3555	97.6334

7.4.5 Discussions

From Figs. 7.4, 7.5, and 7.6 and Tables 7.1, 7.2, and 7.3, we could have three findings. First, no spectrum could compete with all the others for all settings. This is mainly because different light could enhance different features of palms, while these different features have different intensity distributions which are in favor of different parameters.

Second, among the three primary colors, red leads to a little higher accuracy than blue and green. This is mainly because red could not only capture most of the palm line information, but also capture some palm vein structures as shown in Fig. 7.1. This additional palm vein information helps in classifying those palms with similar palm lines. It could also explain why the composite colors (magenta, yellow, white) could get better accuracy than cyan.

Last, white color could not get the best accuracy among the seven spectra. Yellow achieves the best result by the schemes of wide line detection and $(2D)^2$PCA, while magenta achieves the best result by the scheme of competitive coding. Furthermore, there is a statistically significant difference at a significance level of 0.05 between magenta and white in competitive coding trails and between yellow and white in wide line detection trials. This finding could be explained as follows.

According to additive color mixing and object imaging formula (Kittler and Sadeghi 2004), the intensity of palmprint image by white light could be regarded as a composition of three components, intensity by blue light, green light, and red light. As shown in Fig. 7.1, the palmprint images under blue and green illuminations are more similar to each other than to the image under red illumination. A quantitative image similarity index, CW-SSIM (Wang and Simoncelli 2005), further validates that blue and green collect much redundant information, as the average similarity between blue and green is 0.95 (the maximal value is 1), while the average similarity between blue and red, and green and red is 0.83. The redundancy makes white color that fails to capture more information than the yellow or magenta color, and sometimes, the accuracy drops a little. This is also consistent with the finding of Fratric and Ribaric (2008), so fusion of blue and red channels gets better result than fusion of three channels.

7.5 Conclusion

Palmprint recognition has been attracting lots of research attention in the past decade, and many data collection devices have been proposed. In order for good image quality and high data capture speed, using cameras mounted with active lighting sources is the most popular device configuration. Almost all existing devices used white light as the illumination source, but there was no systematic analysis on whether the white light is the optimal light source for palmprint recognition. This chapter made such an effort on answering this problem by establishing a large multispectral palmprint database using our developed device. With the database, we empirically evaluated the recognition accuracies of palmprint images under seven different colors by three different methods. Our experimental results showed that the white color is not the optimal color for palmprint recognition and the yellow or magenta color could achieve higher accuracy than the white color.

However, so far, our data were collected from East Asian residents (more specifically Chinese) only. Since the palm spectral properties of different groups may be different (Angelopoulou 2001), the finding of this work may not valid for other groups. In the future, more samples from other groups will be collected to investigate the best illumination conditions for palmprint recognition.

References

Angelopoulou E (2001) Understanding the color of human skin. In: Proceedings of the SPIE conference on human vision and electronic imaging, vol 4299, pp 243–251 (SPIE)

Connie T, Jin ATB, Ong MGK, Ling DNC (2005) An automated palmprint recognition system. Image Vis Comput 23:501–515

Duta N, Jain AK, Mardia KV (2002) Matching of palmprint. Pattern Recogn Lett 23:477–485

Fratric I, Ribaric S (2008) Colour-based palmprint verification—an experiment. In: The 14th IEEE mediterranean electrotechnical conference, pp 890–895

Guo Z, Zhang D, Zhang L (2009) Is white light the best illumination for palmprint recognition. In: International conference on computer analysis of images and patterns, pp 50–57

Han C (2004) A hand-based personal authentication using a coarse-to-fine strategy. Image Vis Comput 22:909–918

Han Y, Sun Z, Wang F, Tan T (2007) Palmprint recognition under unconstrained scenes. Asian Conf Comput Vis 4844:1–11 (LNCS)

Han D, Guo Z, Zhang D (2008) Multispectral palmprint recognition using wavelet-based image fusion. In: International conference on signal processing, pp 2074–2077

Jain A, Bolle R, Pankanti S (1999) Biometrics: personal identification in network society. Kluwer Academic Publishers, Boston

Kittler J, Sadeghi MT (2004) Physics-based decorrelation of image data for decision level fusion in face verification. Multiple Classifier Syst 3077:354–363 LNCS

Kong A, Zhang D (2004) Competitive coding scheme for palmprint verification. In: International conference on pattern recognition, pp 520–523

Kong A, Zhang D, Kamel M (2009) A survey of palmprint recognition. Pattern Recogn 42:1408–1418

Kumar A, Wong DCM, Shen H, Jain AK (2003) Personal verification using palmprint and hand geometry biometric. In: Proceedings AVBPA, pp 668–675

Lin C, Chuang T, Fan K (2005) Palmprint verification using hierarchical decomposition. Pattern Recogn 38:2639–2652

Liu L (2007) Wide line detector and its applications. Ph. D thesis, the Hong Kong Polytechnic University

Michael GKO, Connie T, Teoh ABJ (2008) Touch-less palm print biometrics: novel design and implementation. Image Vis Comput 26:1551–1560

Ribaric S, Fratric I (2005) A biometric identification system based on Eigenpalm and Eigenfinger features. IEEE Trans Pattern Anal Mach Intell 27:1698–1709

Wang Z, Simoncelli EP (2005) Translation insensitive image similarity in complex wavelet domain. In: IEEE international conference on acoustics, speech and signal processing, pp 573–576

Wang J-G, Yau W-Y, Suwandy A, Sung E (2008) Person recognition by fusing palmprint and palm vein images based on "Laplacianpalm" representation. Pattern Recogn 41:1514–1527

Wong M, Zhang D, Kong W-K, Lu G (2005) Real-time palmprint acquisition system design. IEEE Proc Vis Image Signal Process 152:527–534

Wu J, Qiu Z, Sun D (2008) A hierarchical identification method based on improved hand geometry and regional content feature for low-resolution hand images. Sig Process 88:1447–1460

Zhang D, Shu W (1999) Two novel characteristics in palmprint verification: datum point invariance and line feature matching. Pattern Recogn 32:691–702

Zhang L, Zhang D (2004) Characterization of palmprints by wavelet signatures via directional context modeling. IEEE Trans Syst Man Cybern Part B 34:1335–1347

Zhang D, Zhou Z (2005) (2D) 2PCA: 2-directional 2-dimensional PCA for efficient face representation and recognition. Neurocomputing 69:224–231

Zhang D, Kong W, You J, Wong M (2003) Online Palmprint Identification. IEEE Trans Pattern Anal Mach Intell 25:1041–1050

Zhang L, Zhang L, Zhang D, Zhu HL (2010a) On-line finger-knuckle-print verification for personal authentication. Pattern Recogn 43(7):2560–2571

Zhang D, Guo Z, Lu G, Zhang L, Zuo W (2010b) An online system of multispectral palmprint verification. IEEE Trans Instrum Meas 59:480–490

Zuo W, Zhang D, Wang K (2006) An assembled matrix distance metric for 2DPCA-based image recognition. Pattern Recogn Lett 27:210–216

Chapter 8
Feature Band Selection for Online Multispectral Palmprint Recognition

Abstract Palmprint is a unique and reliable biometric feature with high usability. In the past decades, many palmprint recognition systems have been successfully developed. However, most of previous works use the white light as the illumination source, and the recognition accuracy and anti-spoof capability are limited. Recently, multispectral imaging attracts research attention as it can acquire more discriminative information in a short time. One crucial step in developing online multispectral palmprint systems is how to determine the optimal number of spectral bands and select the most representative bands to build the system. This chapter presents a study on feature band selection by analyzing hyperspectral palmprint data (520–1050 nm). Our experimental results showed that three spectral bands could provide most of discriminate information of palmprint. This finding could be used as the guidance for designing new online multispectral palmprint systems.

Keywords Multispectral palmprint recognition · Clustering · Biometrics · Anti-spoof

8.1 Introduction

Biometric authentication is the study of methods for recognizing humans based on one or more physical or behavioral traits (Jain et al. 1999). As a unique biometric feature, palmprint recognition has been attracting much attention in the past decade (Connie et al. 2005; Hu et al. 2007; Han et al. 2003; Jia et al. 2008). It owns many merits, such as high accuracy, high user-friendliness, and low cost. However, there is much room to improve the palmprint systems, especially in the aspects of accuracy and its vulnerability to spoof attacks (Schukers 2002). Multispectral imaging is a good solution to such improvement (Zhang et al. 2010).

Multispectral imaging could capture a series of palmprint images at various spectral bands simultaneously, recognition accuracy is improved as more discriminative information could be provided, and anti-spoof capability of systems is

D. Zhang et al., *Multispectral Biometrics*,
DOI 10.1007/978-3-319-22485-5_8

enhanced because it is not a trivial issue to make a fake palm having the same spectral signatures with a real palm. Because of the advantages mentioned above, multispectral imaging has been not only applied on palmprint recognition (Zhang et al. 2010; Hao et al. 2007; Rowe et al. 2007; Hao et al. 2008), but also applied on other biometric systems, including multispectral fingerprint recognition (Rowe et al. 2005), multispectral face recognition (Chang et al. 2009; Di et al. 2010), and multispectral iris recognition (Boyce et al. 2006).

There are two underlying key issues need to be addressed well before wide application of multispectral biometrics. First, how many spectra are enough for discriminating different palms? Usually, more feature bands provide more information; thus, higher accuracy could be expected. On the other hand, more feature bands require high cost on feature extraction and matching. Furthermore, because of the redundancy between different spectra, more information may fail to increase the accuracy sometimes (Zhang et al. 2010; Hao et al. 2008). Second, how to choose representative spectra for a given number of feature bands? After determining the number of feature bands, a group of bands could be selected by some rules, such as divergence (Chang et al. 2009), and exhaustive search (Guo et al. 2010). The second issues belong to feature selection (Duda et al. 2001), and some pioneering works have been made (Chang et al. 2009; Guo et al. 2010). However, to our best of knowledge, there is no public report on the first issue for biometric research.

In this chapter, we proposed a clustering-based method to determine the number of feature band from hyperspectral palmprint database. It is found that three spectral bands could contain most of the discriminant information. The finding is also validated by recognition experiments. In the following, Sect. 8.2 introduces the data collection of hyperspectral palmprint cube. Section 8.3 reports the proposed spectral *k*-means clustering algorithm. Section 8.4 shows the validation through verification experiments, and Sect. 8.5 draws conclusion.

8.2 Hyperspectral Palmprint Data Collection

To discover the optimal number of feature band for multispectral palmprint recognition, a hyperspectral imaging system is established. The key components of the system include a liquid crystal tunable filter (LCTF) made by Meadowlark Inc., a charge-coupled device (CCD) made by Cooke Corporation, and two 500-W Osram halogen lights. The full width at half maximum of the filter is 5 nm when the center wavelength is 550 nm. A palm could be imaged at 69 spectral bands with a step length of 10 nm over spectrum of 420–1100 nm. Figure 8.1 shows the imaging system.

After obtaining the hyperspectral cube, a local coordinate of the palmprint image is established (Zhang et la. 2010) in the center band (760 nm), and then, a region of interest (ROI) is cropped from each band based on the local coordinate. Figure 8.2 illustrates the ROIs sample. For the convenience of analysis, these ROIs are

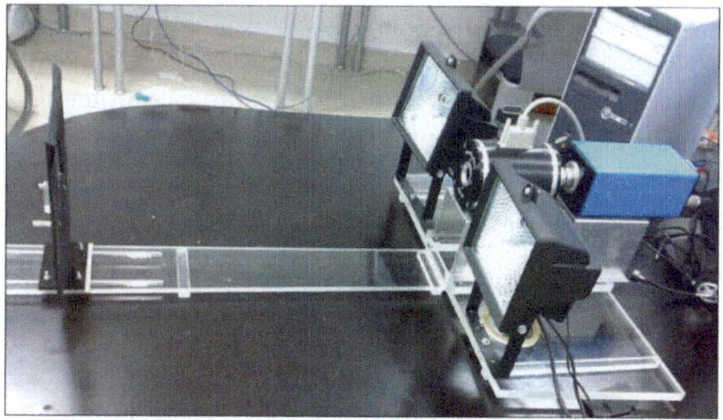

Fig. 8.1 The hyperspectral palmprint imaging system

Fig. 8.2 The ROIs extracted from a hypespectral palmprint cube sample. From *left* to *right*, *top* to *down*, the wavelength is increasing from 420 to 1100 nm with 10 nm interval

normalized to a size of 128 × 128. To reduce the intensity effect, the ROIs are converted to have a mean intensity of 128 with a standard deviation of 20. The first ten and last five spectra are removed due to the low quality of images (Guo et al. 2010). Thus, in the following, only 54 feature bands (520–1050 nm, with 10 nm interval) are used.

A large hyperpsectral palmprint database from 190 individuals is built. The subjects were mainly volunteers from our institutes. In the database, the age distribution is from 20 to 60 years old. The multispectral palmprint cubes were collected by two separate sessions. The average time interval between the two occasions is around 1 month. On each session, the subject was asked to provide around seven cubes of each of his/her left and right palms. So the database contains 5240 images for each band from 380 different palms. Among them, 2608 cubes were collected by the first session, while 2632 by the second session.

8.3 Feature Band Selection by Clustering

k-means clustering algorithm (Duda et al. 2001) is a basic and well-known technique in pattern recognition. During the clustering, some points are clustered into separate classes or clusters according to the given clustering criterion. Figure 8.3 shows an example of k-means. Figure 8.3a shows three clusters which are generated by three separate Gaussian distributions. As k is a critical input parameter, different values will generate different clustering results. Figure 8.3b, d shows the clustering results for $k = 2$, $k = 3$, and $k = 4$, respectively.

When the k is smaller or equal to the true number of clusters, the center distance is very big; while the k is larger than the true number of clusters, one cluster will be

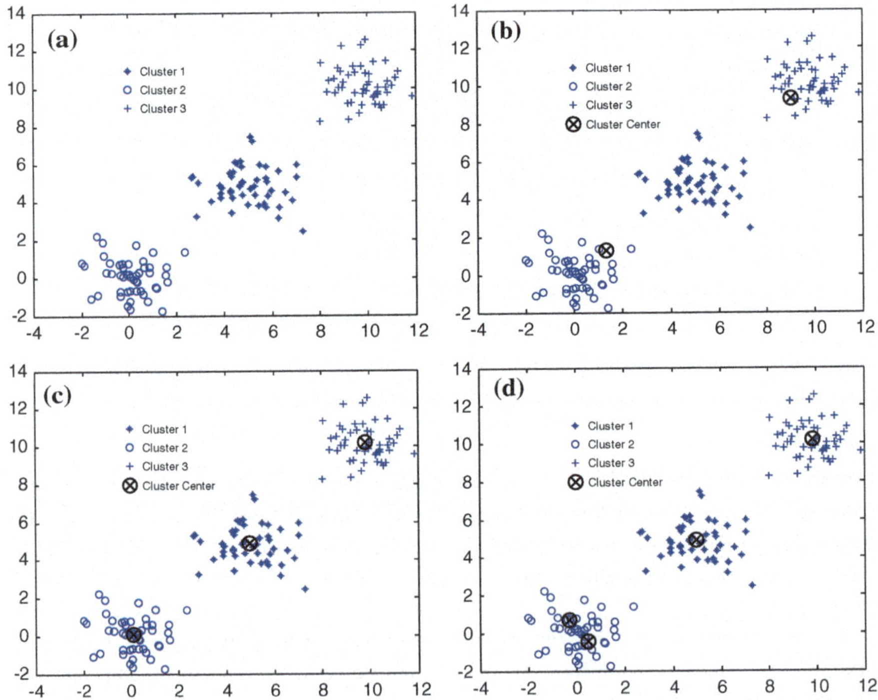

Fig. 8.3 **a** Original data; **b** k-means result, $k = 2$; **c** k-means result, $k = 3$; **d** k-means result $k = 4$

split into small clusters, and the center distance becomes smaller and stable. The minimal center distance is defined as:

$$d_{\min}^{k} = \min_{\substack{1 \leq i \leq k \\ 1 \leq j \leq k \\ i \neq j}} (d_{i,j}) \tag{8.1}$$

$$d_{i,j} = d(c_i, c_j) \tag{8.2}$$

where d is Euclidean distance, c is the cluster center, and k is the number of clusters.

Figure 8.4 shows the $d_{\min}^{k}$ for different k values. As illustrated by Fig. 8.4, the distance drops quickly from k = 2 to k = 3. When k is bigger than 3, the distance changes slowly. This example shows that k-means clustering could be used to discover the true number of clusters. Based on this finding, a spectral k-means clustering algorithm is proposed.

Suppose we have N hyperspectral palmprint cubes with B spectra, as the training set, and Fig. 8.5 shows the pseudo-code of the algorithm:

Here, CW-SSIM distance is used for 2-D palmprint image as it is less sensitive to scale, rotation, and translation variation for palmprint verification (Zhang et al. 2007; Sampat et al. 2009). Where N = 2280 (380 × 6, first six cubes of each palm were selected), B = 54 and T = 100.

Because there is possibility that the clustering ends at local minimal instead of global minimal, for a given k, the clustering runs 1000 times and the most frequent cluster centers are kept as the final result for the given k. Similar as Fig. 8.4, Fig. 8.6 shows minimal center distance for different numbers of clusters.

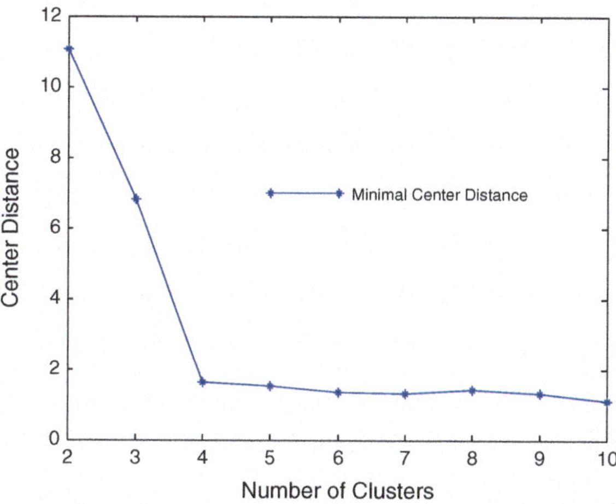

Fig. 8.4 Number of clusters versus minimal center distance

1. Randomly get k initialized centers (wavelengths), x=0;
2. For any wavelength j (j=1,2,…,B), compute the distance between the wavelength with k centers, here the distance is compute as:

$$D_{i,j} = \sum_{n=1}^{N} d_{i,j}^{n} = \sum_{n=1}^{N} d^{n}(c_i, w_j), i = 1,2,...,k$$

where c is the centre wavelength, w_j is the given wavelength, and d is the CW-SSIM distance (Zhang et al. 2007). x=x+1;

3. For each centre, find the wavelengths which have the minimal distance among n centres. There are k clusters of wavelengths:

$$S_i = \{j \mid \arg\min_i D_{i,j} = i\}, i = 1,2,...,k$$

4. For any cluster, find the wavelength which has the minimal average distance with the remaining wavelengths in the cluster, this wavelength is updated as the new center:

$$c_i^{'} = \arg\min_{m \in S_i} \sum_{l=1}^{B} D_{l,m}, l \neq m, l \in S_i$$

5. If the centres are not changed or x=T (T is a predefined iteration threshold), stops; otherwise, goes to step 2.

Fig. 8.5 Pseudo-code of the proposed spectral *k*-means clustering

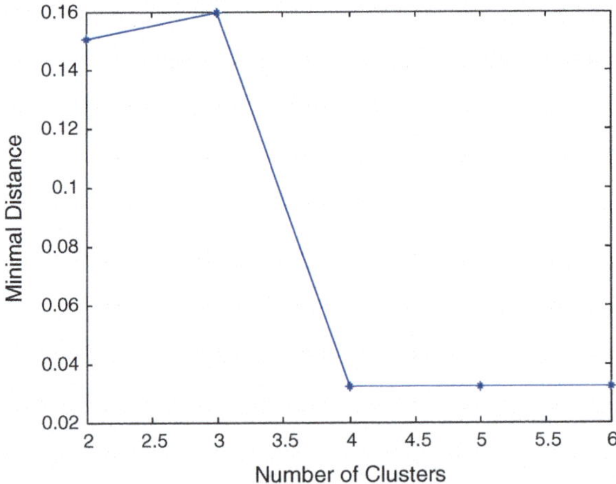

Fig. 8.6 Minimal distance of spectral versus *k*

Figure 8.6 shows that three clusters may be enough to represent the 54 feature bands. The $k = 3$ cluster result is listed in Table 8.1, and the distance map between different wavelengths is plotted in Fig. 8.7. As shown in Fig. 8.7, there are roughly three dark blocks. According to the spectrum definition (Visible spectrum), the three clusters could be roughly named as visible spectrum without red, red spectrum, and near-infrared (NIR) spectrum.

Table 8.1 Clustering result of spectral k-means for $k = 3$

Cluster	Wavelengths
1	520, 530, 540, 550, 560, 570, 580, 590, 600, 610
2	620, 630, 640, 650, 660, 670, 680, 690, 700, 710, 720, 730, 740, 750, 760, 770, 780
3	790, 800, 810, 820, 830, 840, 850, 860, 870, 880, 890, 900, 910, 920, 930, 940, 950, 960, 970, 980, 990, 1000, 1010, 1020, 1030, 1040, 1050

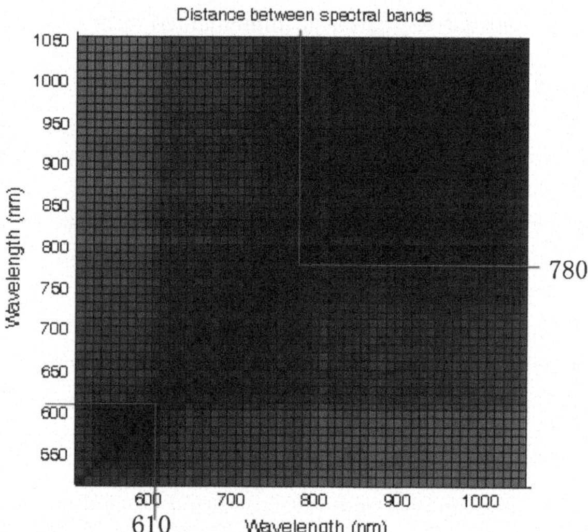

Fig. 8.7 Distance map between different feature bands. The distance is converted from [0, 1] to [0, 255] for display

8.4 Clustering Validation by Verification Test

To demonstrate whether three is the optimal number of feature band, verification experiments are implemented to validate it. Here, CW-SSIM-based verification is used as it is less sensitive to scale, rotation, and translation variation for palmprint verification (Zhang et al. 2007; Sampat et al. 2009).

One cube of each palm in the first session is randomly selected as the gallery sample, and all the cubes in the second session are used as probe samples. For any feature band, the image distance between each sample in the gallery set and sample in the probe set is computed. Totally, there are 1,000,160 (380 × 2632) matching distances for each feature band. Among them, 2632 distances are genuine matching distances, while the remaining are impostor matching distances. To get unbiased results with training gallery selection, the gallery set is selected three times independently. Then, one cube of each palm in the second session is randomly selected as the gallery sample, and all the cubes in the first session are used as probe samples. We use the same test protocol for each wavelength and get 991,040 (380 × 2608) matching distances for each feature band. Among them, 2608 distances are genuine matching distances. Similarly, the gallery set is selected three times independently. Thus, totally six verification trials are implemented. Equal error rate (EER, when false acceptance rate (FAR) equals to false rejection rate (FRR)) is used to evaluate the accuracy. In the following, the average of the six trials is listed as experimental results. Figure 8.8 illustrates the EER with different wavelengths. The lowest EER is achieved by 790 nm with EER = 0.2220 %.

To find the optimal combination for a given number of feature bands, many algorithms could be used to search for the combination, such as divergence (Chang et al. 2008), mutual information (Guo et al. 2006), and entropy (Wang et al. 2006). Here, exhaustive search is implemented as it has less possibility to miss the true optimal combination. Sum score-level fusion (Ross et al. 2006) is used as the fusion technique on the dataset. The lowest EER for each combination is listed in Table 8.2.

Table 8.2 shows more feature bands and higher accuracy. For example, fusion of two bands could reduce the EER from 0.222 to 0.1039 %. However, the improvement from three bands to four bands is very small; it is only 0.0053 %. Using statistical analysis (Mendenhall et al. 2003), the difference between fusion of

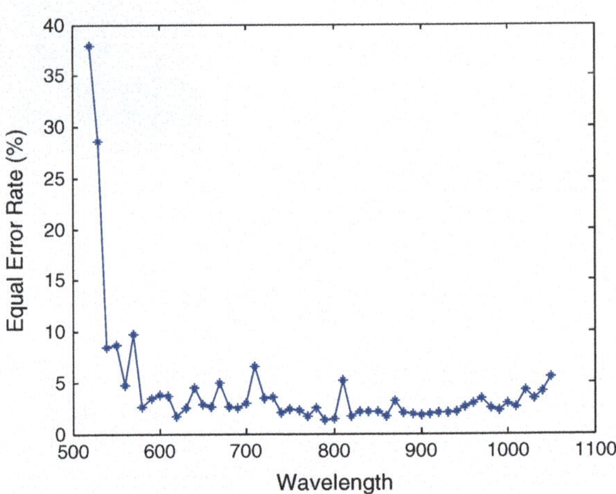

Fig. 8.8 Wavelength versus EER

Table 8.2 Fusion results for different number of feature bands

Number of feature bands	Optimal combination (nm)	EER (%) (Mean $\pm$ standard deviation)
1	790	0.2220 $\pm$ 0.0652
2	580, 770	0.1039 $\pm$ 0.0431
3	580, 760, 990	0.0780 $\pm$ 0.0344
4	580, 620, 760, 940	0.0727 $\pm$ 0.0430

four bands and three bands is not statistically significant. And as shown in Table 8.1, 580, 760, and 990 nm come from three different clusters. These findings are consistent with the results in the above section.

8.5 Summary

In this chapter, a spectral *k*-means clustering algorithm is proposed to cluster hyperspectral palmprint cubes. The clustering could be used to determine an optimal number of feature bands. The result shows that three feature bands may be enough to represent the palmprint features. Based on score-level fusion and exhaustive search on all fusion candidates, three feature bands could get much better results than two bands and it can get comparable results with four bands. This finding validates the effectiveness of the proposed clustering algorithm, and it is empirically demonstrated that three feature bands is a good option for real multispectral palmprint applications. In the future, a new online multispectral palmprint system will be designed based on the finding in this chapter.

However, the hyperspectral database is limited to Chinese persons only, whether the finding is applicable to other groups needs further investigation.

References

Boyce C, Ross A, Monaco M, Hornak L, Li X (2006) Multispectral iris analysis: a preliminary study. In: IEEE computer society conference on computer vision and pattern recognition. Workshops, pp 51–59

Chang H, Yao Y, Koschan A, Abidi B, Abidi M (2008) Spectral range selection for face recognition under various illuminations. In: International conference on image processing, pp 2756–2759

Chang H, Yao Y, Koschan A, Abidi B, Abidi M (2009) Improving face recognition via narrowband spectral range selection using Jeffrey divergence. IEEE Trans Inf Forensics Secur 4(1):111–122

Connie T, Andrew T, Goh K (2005) An automated palmprint recognition system. Image Vis Comput 23:501–505

Di W, Zhang L, Zhang D, Pan Q (2010) Studies on hyperspectral face recognition in visible spectrum with feature band selection. IEEE Trans Syst Man Cyberns Part A Syst Hum 40:1354—1361

Duda RO, Hart PE, Stork DG (2001) Pattern classification. Wiley, Berlin

Guo B, Gunn SR, Damper RI, Nelson JDB (2006) Band selection for hyperspectral image classification using mutual information. IEEE Geosci Remote Sens Lett 3:522–526

Guo Z, Zhang L, Zhang D (2010) Feature band selection for multispectral palmprint recognition. In: International conference on pattern recognition, pp 1136–1139

Han C, Cheng H, Lin C, Fan K (2003) Personal authentication using palm-print features. Pattern Recogn 36:371–381

Hao Y, Sun Z, Tan T (2007) Comparative studies on multispectral palm image fusion for biometrics. In: Asian conference on computer vision, pp 12–21

Hao Y, Sun Z, Tan T, Ren C (2008) Multispectral palm image fusion for accurate contact-free palmprint recognition. In: International conference on image processing, pp 281–284
Hu D, Feng G, Zhou Z (2007) Two-dimensional locality preserving projections (2DLPP) with its application to palmprint recognition. Pattern Recogn 40:339–342
Jain A, Bolle R, Pankanti S (1999) Biometrics: personal identification in network society. Kluwer, Boston
Jia W, Huang D, Zhang D (2008) Palmprint verification based on robust line orientation code. Pattern Recogn 41:1504–1513
Mendenhall W, Beaver RJ, Beaver BM (2003) Probability and statistics. Thomson, Brooks/Cole
Ross AA, Nadakumar K, Jain AK (2006) Handbook of multibiometrics, Springer, Berlin
Rowe RK, Nixon KA, Corcoran SP (2005) Multi spectral fingerprint biometrics. In: Proceedings of information assurance workshop, pp 14–20
Rowe RK, Uludag U, Demirkus M, Parthasaradhi S, Jain AK (2007) A multispectral whole-hand biometric authentication system. In: Proceedings of biometric symposium. Biometric consortium conference, pp 1–6
Sampat MP, Wang Z, Gupta S, Bovik AC, Markey MK (2009) Complex wavelet structural similarity: a new image similarity index. IEEE Trans Image Process 18:2385–2401
Schukers SAC (2002) Spoofing and anti-spoofing measures. Inf Secur Tech Report 7:56–62
Wang H, Angelopoulou E (2006) Sensor band selection for multispectral imaging via average normalized information. J Real-Time Image Proc 1:109–121
Zhang L, Guo Z, Wang Z, Zhang D (2007) Palmprint verification using complex wavelet transform. In: International conference on image processing, pp 417–420
Zhang D, Guo Z, Lu G, Zhang L, Zuo W (2010) An online system of multi-spectral palmprint verification. IEEE Trans Instrum Meas 59:480–490

Part IV
Multispectral Hand Dorsal Recognition

Chapter 9
Dorsal Hand Recognition

Abstract Dorsal hand recognition has drawn much attention due to its commonality, uniqueness, and stability in biometrics. In particular, the hand vein, which is considered as the main identification information in dorsal hand, is a living feature and hard to be fabricated, so it reflects the huge advantage in improving the anti-spoof ability. However, the imaging of subcutaneous feature needs more requirements for the light source than other biometric features such as face. Unsuitable light source would result in poor-quality dorsal hand image with amount of important information loss or be mixed with useless information. To address the problem of spectra selection is a very effective way to improve final recognition performance of dorsal hand. Multispectral technique is just the one that helps us to search for the optimal band for light source around visible light part and near-infrared (NIR) light part in image acquisition process. On the other hand, the feature pattern is one of the key factors influencing the recognition performance. Considering that multispectral analysis should be implemented on the same feature pattern overall the whole spectra, applying a proper one is necessary for making sure that it can be suitable for not only dorsal hand vein extraction, but also multispectral images.

Keywords Dorsal hand recognition · Multispectral image · Band selection · Feature estimation

9.1 Introduction

With human's more attention on security issue, security technologies have become increasingly demanding. Conventional biometric features such as face and fingerprint have been developed to a high level with enough precision for large-scale database. Researchers have changed their focus on anti-counterfeiting gradually during the past decade. Combining more biometric features together to set up multimodal identification system is a feasible solution to increase the difficulty of

D. Zhang et al., *Multispectral Biometrics*,
DOI 10.1007/978-3-319-22485-5_9

counterfeiting. Nevertheless, seeking for a living feature which is also easy for acquisition and processing can improve the security property effectively. Dorsal hand vein is just the one fitting with the demand of forgery prevention as a sub-application of biometrics. The vein presents a kind of reticular formation under dermis which provides high anti-destroyed ability.

Dorsal hand recognition is a technical method providing identity authentication on the basis of vein information difference. Accurate vein extraction has always been the main topic throughout the development history of this field. Early studies started to seek for better means of vein segmentation in the 1990s (Im et al. 2001). Then, the vein-based feature representation method with the process of vein thinning and skeletonization led the main developing trend in the flowing ten years (Kumar and Prathyusha 2009; Ding et al. 2005; Li et al. 2010a, b). The related research work revealed that the performance of dorsal hand vein is good enough for its being a comparative alternative to traditional biometric model such as fingerprint and palmprint. Meanwhile, dorsal hand is also taken as supplemental information in the fusion with other biometric features. Aycan Yuksel claimed that EER of palm–dorsal fusion scheme can reduce from 1.43 % to zero compared with single palm module on a database of 70 subjects (Yuksel et al. 2010).

Besides the distance calculation for extracted vein feature or classifier design and selection, the vein extraction quality is the most important factor that influences the final dorsal hand recognition performance. The high quality of vein extraction is attributed to not only proper preprocessing methods and accurate segmentation, but also original high-quality image. Due to the reason that the vein is a kind of subdermis information, the imaging quality in acquisition procedure is commonly not as high as other biometric features. The collected dorsal hand image may run into problems such as broken and blurred vein, so image quality has become a bottleneck element restricting further development of dorsal hand recognition. Limited to device and technique reasons, optimizing light source to obtain clear and strong contrast image is always an ignored issue for studying.

Far-infrared (FIR) source (8–14 μm) and near-infrared (NIR) source (700–1000 nm) are two widely used noninvasive imaging technique for dorsal hand. FIR is sensitive to ambient temperature and humidity conditions, even the state of human body can also result in unstable image quality. On the other side, NIR may meet the problem of vein disruption due to the interference of visible skin features such as fine hairs, scar, or other line patterns. Lingyu Wang first initialed a system to test the performance difference between FIR and NIR (Wang et al. 2007); the two imaging methods can both identify all the subjects correctly. Unfortunately, the result appears not to be convincing enough as the database is too small. However, he preferred NIR because of its higher stability.

According to NIR imaging theory, deoxyhemoglobin in blood vessel can absorb more NIR, so the vein part appears to be darker than the surrounding part. The absorptivity difference leads to gray difference on dorsal hand image and forms clear vein structure. When the wavelength of light source changes, the absorptivity coefficients of vein and non-vein parts also change and result in texture and contrast change on collected image. Even if considering the vein part only, the change is not uniform due to the variety of vein size, epidermal thickness, and pigment influence. High-quality dorsal hand image should meet these conditions: clear vein boundary, continuous vein line, and less noise. Light source can no doubt change the image quality, and find out the optimal band which will have the practical meaning of recognition rate improvement.

Optimal band selection for biometrics is an emerging tool to improve recognition performance with the help of multispectral technique these years. The research about this field was in the blank stage in the past. Most dorsal hand acquisition systems use 850 nm as the light source to construct LED array lamps. It is considered that the image quality can reach the optimal state by visual observation under 850 nm and this band is widely adopted as ready-made experience (Wang et al. 2008a, b; Cheng et al. 2007), though 850 nm does not totally match 760 nm, around which band deoxyhemoglobin in blood vessel has the highest absorptivity rate according to biomedical research (Matcher et al. 1995). This method can be classified into image quality estimation, and a more objective criterion should be established to replace the conditional visual observation-based method. Another general way of band selection is to follow the characteristics of the acquisition system. Sanchit chooses 830 nm as the work band, because the spectral response curve of the camera's NIR sensor can peak at this band (Sanchit et al. 2011). Putting undue emphasis on system's transmission characteristics must neglect the action of dorsal hand's absorptivity mechanism. Even if the transmission and absorption curve are combined together, it is still a much less straightforward task than image quality estimation. Information entropy, contrast measure, PCA ranking, and frequency measure are several kinds of quality estimation in different views. All these quality estimation measurements try to construct a linkage between their one-sided criteria and the final performance and further obtain the optimal result through ranking. Traditionally, such methods refer to unsupervised method, as opposed to supervised method. The disadvantage of unsupervised method is apparent that the ranking result is not absolute accuracy and we usually just get suboptimal result for easy operation purpose. Obviously, a more authentic way of band selection is to follow a recognition rate-based method which belongs to supervised method.

9.2 Multispectral Acquisition System and Database

Besides adoption variety of different parts on dorsal hand, light with various wavelengths also has different capability of penetrating dorsal hand skin. So the main layer of the texture illuminated is different. For example, the spectra with longer wavelength can reach deeper skin layer and the small vein can be recorded clearly. On the other side, the bands near visible light part focus more on surface skin and the vein pattern would get blurred. In fact, dorsal hand image has a process of slow change, while the wavelength decreases with fixed step. Companied by surface skin, the vein is reflected in the whole NIR spectra range, and only the ratio of information possessed by each part is different. The hybrid situation causes image blurring and becomes a disadvantage to other biometrical features acquired under visible light. There does not exist a sharp line dividing the spectrum into two parts which reflect vein and skin surface, respectively. To investigate the skin surface texture's negative influence or explore its positive distinguishing capability, part of visible light needs to be taken into consideration for the task of optimal single band selection. So it is necessary to make use of multispectral technique and set up an image acquisition system to realize continuous sampling ranging across both visible spectra and NIR spectra.

9.2.1 Image Acquisition System

To the best of our knowledge, there is no currently publicly available dorsal hand database including both visible and NIR spectra simultaneously. In order to study the optimal band in such a large spectrum region, establishing a new dorsal hand database with a wide range of utility is necessary. It can also offer the scope for us to study the bands of visible and NIR separately.

Halogen lamps are chosen as the light source for our multispectral acquisition system, which is built on the foundation of reflective imaging principle. The system structure drawing is shown in Fig. 9.1a. A widely taken approach of getting dorsal hand image is to connect the optical filter and monochrome sensor together (Wu et al. 2010). In our designed system, the reflected light of dorsal hand is filtered into a narrow band after passing through the liquid crystal tunable filter (LCTF) made by Meadowlark Inc. With the help of adjustable lens, the filtered light can be focused on the CCD camera. In the process of image collecting, volunteers are asked to clench fist and put back of their hands in front of LCTF. To obtain good

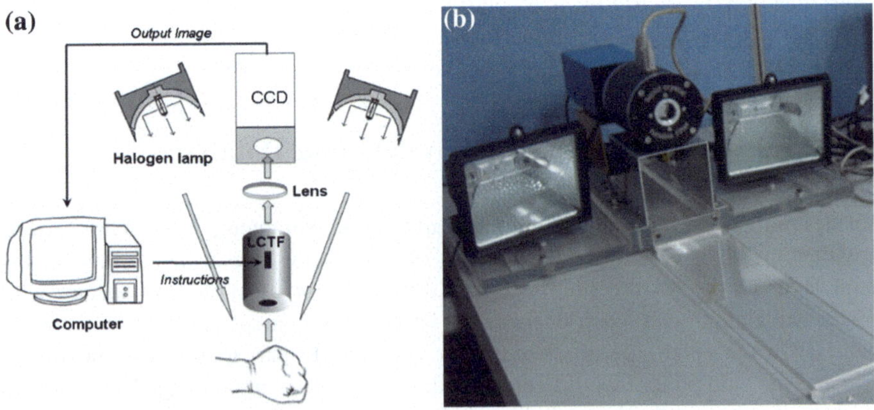

Fig. 9.1 The structure and physical representation of multispectral capture system

quality image, a special hand attachment is designed to make the dorsal hand flat enough and perpendicular to the lens, with rotation angle as small as possible. The computer sends instructions to LCTF automatically so that the band allowed to pass can realize fast switching and get into new stable state no more than 100 ms.

The optical filter equipment LCTF has good transmission performance within a large spectrum region from 400 to 1100 nm, including both visible light and NIR light. Due to the character that the half-power bandwidth is less than 5, 10 nm is chosen as the band interval for multispectral images. As the light source power radiation from the halogen lamp is centered on the spectra 520–1040 nm, the multispectral capture system mainly covers these 53 bands, not only near-infrared light (700–1040 nm), but also visible light with the spectra of green (520–580 nm), yellow (580–600 nm), orange (600–620 nm), and red (620–700 nm) components.

The original collected multispectral dorsal hand database is composed of 211 different individuals including both their left hands and right hands. The volunteers are all East Asians with ages ranging from 20 to 50, and the majority of them are young people. Every subject contains 5 groups of multispectral images, each of which has a total of 53 single bands. The collected samples are all 512 × 501 pixels with a gray-scale resolution of 8 bit per pixel.

The vein patterns of left and right hands are not totally bilateral symmetrical, including structure deviation and location deviation, especially for small blood vessels. It is assumed that the dissimilarity is big enough for us to treat the left and right hands of the same volunteer as two from different ones (Badawi 2006). So the multispectral dorsal hand database can be regarded to own 422 different hands for identification test. Figure 9.2 illustrates a volunteer's left and right dorsal hand images under several bands.

9.2.2 ROI Database

Because common dorsal hand matching methods are almost based on spatial domain, it places a higher demand on the database for less translation and rotation deviation. Constructing a region of interest (ROI) database directly can bring convenience for further matching task. In this process, we make sure that the dorsal hand information can be retained to the maximum extend while removing the background.

ROI extraction resorts to good objection detection and localization method. Followed by Gaussian smoothing with a template of 4 × 4 size, Ostu thresholding is chosen as a binarization tool to separate hand and background (Ostu 1979). The key point of this foreground extraction method is to optimize the threshold to maximize the variance between two classes. From Fig. 9.3b, we can see that the hand part in the binary image may have many empty holes and narrow connection parts with other interferential parts together after binarization. The two troubles can be dealt with by morphological open–close operations. Meanwhile, the boundary of hand part can get smoother. The largest connected area in the binary image is selected as

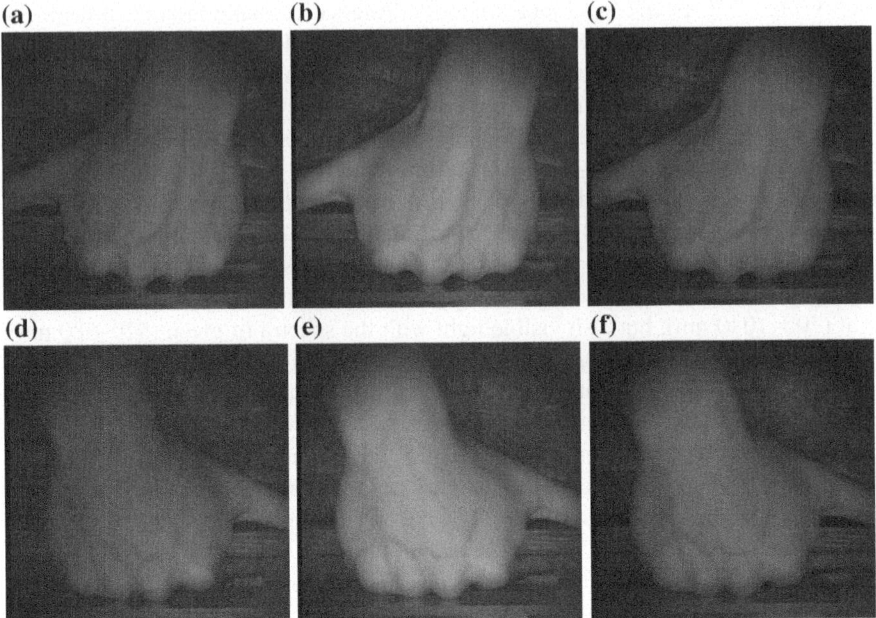

Fig. 9.2 Multispectral images of *left* and *right* dorsal hands all from the same person. **a** Left hand under 650 nm. **b** Left hand under 800 nm. **c** Left hand under 950 nm. **d** Right hand under 650 nm. **e** Right hand under 800 nm. **f** Right hand under 950 nm

the dorsal hand part, and then, the outline of dorsal hand should be recorded with a proper boundary searching methodology to realize precise positioning.

By observing morphological operation result in Fig. 9.3c, the lower part of dorsal hand usually has more clear borders and contains enough information for determining the position and rotation angle of whole hand. So only the lower part of hand outline needs to be extracted. First, a predefined horizontal line is drawn in an

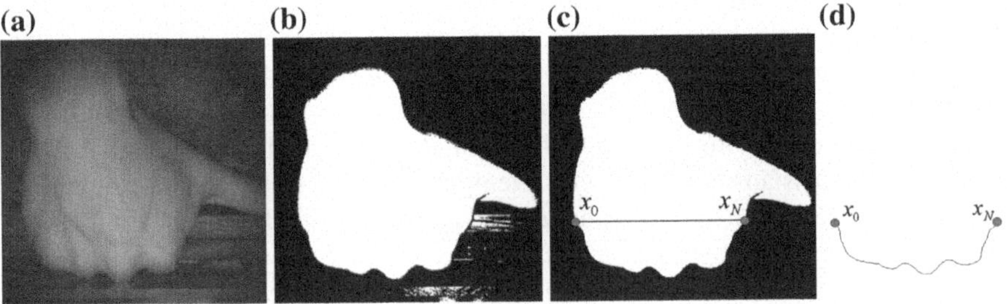

Fig. 9.3 The result in each step of boundary searching. **a** The original dorsal hand image. **b** Binarization result with smoothing and OSTU. **c** The largest connected area with predefined *horizontal line*. **d** The boundary of dorsal hand's lower part

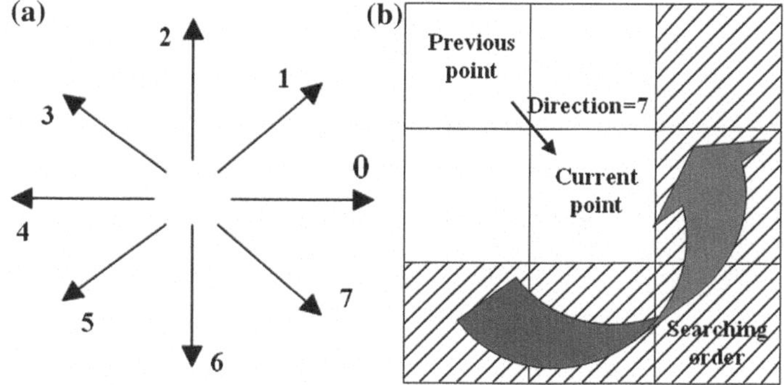

Fig. 9.4 **a** Definition of chain direction. **b** Boundary search in each step; the *shadow* presents 5-candidate next boundary point

appropriate height to get left and right end points x_0 and x_N, and N is the total point number on lower part's boundary. With the help of 8-connected boundary tracking algorithm, one-pixel-width dorsal hand boundary is recorded continuously.

This algorithm utilizes the conception of direction chain code to search for the next boundary point serially and construct a boundary vector from x_0 to x_N in sequence. For each current point, directions should be indexed from 0 to 7 like Fig. 9.4a shows. The boundary vector is denoted by $B = \{x_0\}$ initially. We define chain direction as position relationship between current point and previous point. As the largest connected dorsal hand part is 4-connected, once the direction is definite, probable position of next point could only be in five neighboring positions, which can form a semicircle as illustrated in Fig. 9.4b. If the chain direction changes, the five probable next positions rotate around the current point with corresponding angle. Because the foreground pixels are all denoted by 1 in binary image, the position of the first occurrence of 1 in counterclockwise order is seen as the next point on the boundary. The detailed iterative steps are displayed as follows:

Step 1: Take point x_0 as the current point. Initial the chain direction to be 7.

Step 2: Find the first occurrence neighboring position of foreground from five candidate places determined by direction and denote it by x_n, $B = \{B, x_n\}$.

Step 3: Renew the current point and chain direction and return back to Step 2 until $x_n = x_N$.

As the volunteers are asked to clench their fists with the purpose of stretching the dorsal hand skin, the invariant feature (IF) points on the dorsal hand boundary can be extracted with the similar method as Yuan refers in his work (Yuan et al. 2010). A point M is labeled in the middle position of the line segment connecting x_0 and x_N. Then, the distances from M to boundary points are calculated in sequence to get

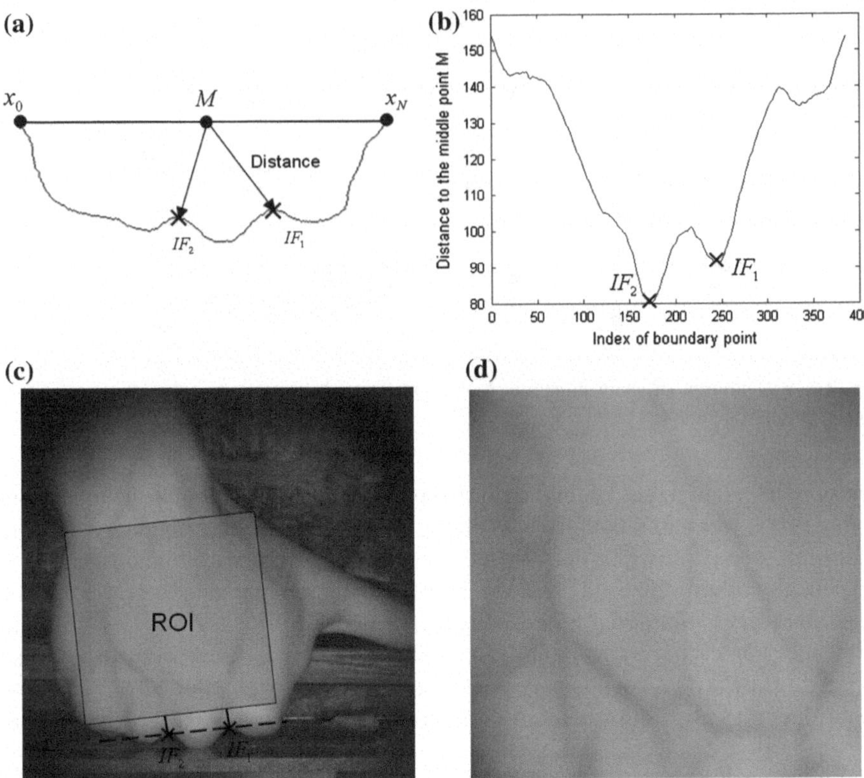

Fig. 9.5 An illustration of ROI extraction on 800 nm. **a** The distance from middle point to boundary points. **b** Search IFs based on local minimum value. **c** Region extraction based on two IF points. **d** ROI image after rotation

the distance curve as shown in Fig. 9.5b. In general, two local minimal points IF_1 and IF_2 on the distance curve can be easily extracted out. They represent the concave point between index finger and middle finger, and the concave point between middle finger and right finger, respectively. To ascertain dorsal hand orientation, a new coordinate system needs to be established along the straight line connecting the two IF points. Then, move the line upward along the orthogonal direction in the new coordinate system and take a square to cover the original image for ROI extraction. A maximum fixed size of the square is chosen while guaranteeing that the proportion of valid dorsal hand part to the whole ROI image can still keep the value near 100 %. The statistical result shows that 224 × 224 is the optimal size for ROI image.

9.3 Feature Representation

9.3.1 Introduction of Dorsal Hand Feature Representation

The final task is to seek for the optimal single band which can contribute most to the dorsal hand recognition. The feature pattern is an instable factor influencing the final judgment. Different feature extraction approaches extract distinguishable information with different emphasis, and the performance would vary slightly even on the same band. We expect to sum up current mainstream methods of dorsal hand feature representation and propose convincing criterion for feature pattern selection. So the band selection task is partly converted to the problem of feature pattern comparing.

For dorsal hand recognition, the first widely adopted way of feature representation is the vein method based on which the structural information is further extracted for recognition. The whole procedure is illustrated in Fig. 9.6 with three parts. Threshold segmentation method is firstly used to extract the vein part and generate a corresponding binary image. Then, different thinning algorithms followed by pruning process if necessary are applied to obtain one-pixel-width skeleton of the dorsal hand vein. At last, the minutiae including the crossing and end points are extracted for further distance calculation. Point-to-point Euclidean distance, Hu invariants, and modified Hausdorff distance are testified to have high distinguishing capability through calculating and comparing spatial distance between minutiae sets by Liu et al. (2004), Ding et al. (2005), and Li et al. (2010a, b).

Structural information has good property of being insensitive to translation and rotation, and it is very suitable for nonalignment ROI image. But in Wang's work (Wang et al. 2008a, b), he also analyzed its tolerance to minutiae missing and pointed out that the increasing number of missed minutiae would cause dramatic rise of EER. Generally speaking, the minutiae number out of dorsal hand vein is not more than 15, so several minutiae missing would lead to serious mistake in genuine matching. Structural method is more suitable for high-quality image with clear vein structure. Obviously, our multispectral database cannot meet this requirement, because the sharpness of vein is not invariant across the whole spectra. No matter we use any kinds of distance measure based on vein skeleton image, the fragmentary and blurred veins would no doubt cause remarkable error. Meanwhile, the thinning procedure equals to making a simplification on the vein's location information, and it also loses much important information like the width of vein. So reserving all foreground pixels for matching is more reliable for multispectral

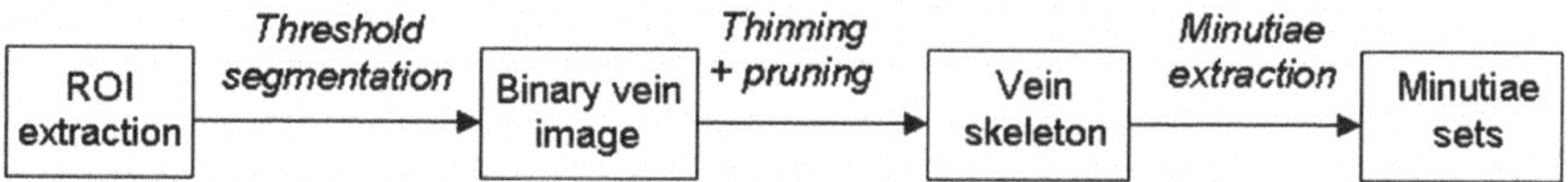

Fig. 9.6 The process of vein extraction based on structure representation

database. Within vein-based method, we tend to use line-based method to generate more complete vein shape instead of structure-based method.

On the other side, the contribution of vein and skin texture to dorsal hand recognition has the chance to be confirmed with the assistance of visible–NIR database. Whether skin texture under visible light can play a comparable role as vein is expected to be testified. So the extraction of an effective and robust feature not limited to conventional dorsal hand vein is required, and it should be verified by synthesis comparison of several feature patterns depending on different recognition scheme such as the appearance-based method and coding-based method. $(2D)^2$PCA (Zuo et al. 2006) and Gabor filter (Zhang et al. 2003) are two mature methodologies in palmprint recognition on the foundation of taking the whole image and texture information as research objective. Gabor filter method further develops to a new form for higher stability and performance combined with a coding method called CompCode (Kong and Zhang 2004). Here, we apply them and try to analyze the applicability of them in multispectral dorsal hands.

9.3.2 $(2D)^2$PCA

The strategy taken by $(2D)^2$PCA is to train a group of samples to get statistical projector parameters for matching. It can be seen as a two-dimensional extended technique of PCA for matrix data like images. This subspace learning method is prior to PCA in terms of facility, as it can construct covariance matrix directly from original images instead of transforming them into vectors. By contrast, the reduction of dimension makes $(2D)^2$PCA to be more appropriate for small sample size database computationally convenient than PCA. It has also been testified that the recognition accuracy of $(2D)^2$PCA is higher (Yang et al. 2004). This method mainly studies the pattern for each pixel based on the samples of the same class. It can be regarded as a common methodology for biometrics in spite of the texture expression characteristics.

$(2D)^2$PCA constructs row and column projectors from two covariance matrices. Take the row covariance matrix for example; an image with size of $M \times N$ is treated as an M-set of $1 \times N$ row vector. The row and column covariance matrices are generated following the formula below:

$$C_r = \frac{1}{K}\sum_{k=1}^{K} (X_k - \overline{X})^T (X_k - \overline{X}) \tag{9.1}$$

$$C_c = \frac{1}{K}\sum_{k=1}^{K} (X_k - \overline{X}) (X_k - \overline{X})^T \tag{9.2}$$

Suppose $X_k(k = 1, 2, \ldots, K)$ is a sample of certain band. $\overline{X}$ is the mean matrix of K training samples. The projector matrix $V_r = \left[v_r^1, v_r^2, \ldots, v_r^{k_r}\right]$ and $V_c =$

$\left[v_c^1, v_c^2, \ldots, v_c^{k_c}\right]$ $(k_c \ll M, k_r \ll N)$ consists of a subset of eigenvectors corresponding to several orthogonal coordinates, the cumulative energy of which can account for the majority of covariance matrix.

Once the projector matrix is obtained, any pair of images for comparison can be converted to a new coordinate system with orthogonal linear transformation. The distance between gallery image G and probe image P is defined as follows:

$$d_P = \left\| V_c^T G V_r - V_c^T P V_r \right\| \tag{9.3}$$

9.3.3 CompCode

2-D Gabor phase coding scheme is primarily explored for palmprint recognition to do the matching relying largely on the principle lines. This approach can manage the low-quality or resolution image without clear wrinkles to the maximum advantage. Daugman proposed a circular Gabor filter as an effective tool for texture analysis (Daugman 1993). If the spatial aspect ratio is added, the Gabor filter can be converted to ellipse shape. So a more general form of Gabor filter can be written as follows:

$$\begin{cases} G(x, y, \sigma, \omega, \theta, \varphi, \gamma) = \exp\left(-\dfrac{(\gamma x')^2 + y'^2}{2\sigma^2}\right) \exp(i(\omega x' + \varphi)) \\ x' = x\cos\theta + y\sin\theta \\ y' = -x\sin\theta + y\cos\theta \end{cases} \tag{9.4}$$

where ω is the angular velocity of sinusoidal wave, θ represents the orientation of the mask, φ represents the phase offset, and γ is the spatial aspect ratio. The Gabor filter can be confined to a finite area with restricted conditions like $(\gamma x')^2 + y'^2 \leq 3\sigma$. According to the number of orientation N_o, which is commonly an even number, θ is assigned with the value $\pi, \pi/N_o, \ldots, (N_o - 1)\pi/N_o$. A total of $2N_o$ masks are used to get the convolution values based on the real part and imaginary part of Gabor filter, and the orientation corresponding to the maximum one is used to relabel each pixel with $n(n = 0, 1, \ldots, N_o - 1)$.

Assume that G_R and P_R are two compared real parts of Gabor feature matrix with the size $M \times N$, a similarity measurement called normalized hamming distance is a good tool for code matching. For more convenience in matching process, the new labeled direction image can be coded with competitive coding (CompCode) scheme (Kong and Zhang 2004). $N_o/2$ bits are needed for storage of each pixel, and each Gabor feature images can be split into $N_o/2$ binary images.

In order to alleviate the mismatching problem due to the position deviation, pixel-to-area matching scheme is introduced to the normalized hamming distance calculation (Jia et al. 2008). Formula 9.5 explains the calculation in detail, and $\otimes$ represents XOR operator. The distance from G to P $d_C(G, P)$ sums up the minimal distance between each pixel in feature matrix G_R and its corresponding neighboring area in P_R. U and V mean the tolerated position deviation, and they determine the size of neighboring area.

$$d_C(G,P) = \frac{\sum_{i=1+U}^{M-U} \sum_{j=1+V}^{N-V} \min_{|u| \leq U, |v| \leq V} \left\{ \sum_{k=1}^{N_o/2} G_R^k(i,j) \otimes P_R^k(i+u,j+v) \right\}}{(N_o/2)\,(M-2U)\,(N-2V)} \tag{9.5}$$

Formula 9.5 is directional, and we can get the distance $d_C(P,G)$ from image P to G in similar way. The final distance is the smaller one out of the two values.

9.3.4 *MFRAT*

Line feature extraction is a vein-targeted method. It treats the vein as the main foreground and needs designing proper parameters to obtain binary feature image. It is observed that the gray value of vein is obviously lower than other parts and the cross section of vein appears to be an inverted bell shape. A widely used method of this kind in biometrics is called modified finite radon transform (MFRAT), and it presents its effectiveness for principle line detection depending on the flexibility to defining the line's width and length (Jia et al. 2008). MFRAT cumulates the image pixels over a set of lines, and dorsal hand vein can be seen as straight line approximately within a limit length.

$$r(x_0, y_0, \theta) = \sum_{w=-W}^{W} \sum_{x=-(L-1)/2}^{(L-1)/2} \sum_{y=-(L-1)/2}^{(L-1)/2} f(x_0+x, y_0+y)\ \delta(y - x\tan\theta = w) \tag{9.6}$$

The formula above expresses the MFRAT value calculation for each pixel (x_0, y_0) based on certain orientations θ. The width of MFRAT templates is $2W+1$, and L represents the length. Unlike CompCode scheme, the energy map is utilized in place of direction map. Only, the minimum amplitudes of MFRAT value out of all directions are recorded for each pixel. Given a ROI image after image enhancement, we can get the line feature image following these steps of MFRAT.

1. Subtract the mean value of whole image from input image.
2. Calculate integral value along different directions for each pixel.
3. Choose the minimum one as the energy value.
4. Get line feature map after binarization.

Templates with 12 orientations $(0, \pi/12, \pi/6, \ldots, 11\pi/12)$ are applied on Radon transform to replace the energy image calculation in Formula 9.6. The template size

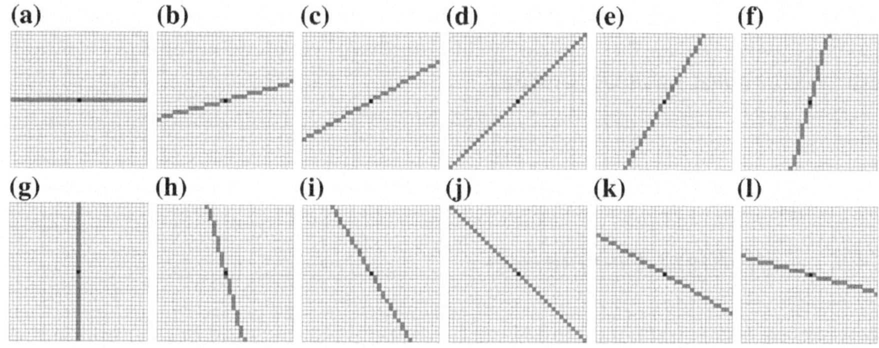

Fig. 9.7 Templates used in MRFT with 12 directions. **a** $\theta = 0$; **b** $\theta = \pi/12$; **c** $\theta = \pi/6$; **d** $\theta = \pi/4$; **e** $\theta = \pi/3$; **f** $\theta = 5\pi/12$; **g** $\theta = \pi/2$; **h** $\theta = 7\pi/12$; **i** $\theta = 2\pi/3$; **j** $\theta = 3\pi/4$; **k** $\theta = 5\pi/6$; **l** $\theta = 11\pi/12$

is 35 × 35 pixels; the value is adjusted to our multispectral dorsal hand database. Figure 9.7 is an illustration of 12 orientation MFRAT templates.

Given the binary form of line feature image F_G and F_P, the similarity score can be calculated with pixel-to-area scheme. We denote the distance from F_G to F_P by

$$d_M(F_G, F_P) = \frac{\sum_{i=1+U}^{M-U} \sum_{j=1+V}^{N-V} \min_{|u| \le U, |v| \le V} \{(F_G(i,j) \otimes F_P(i+u, j+v)) \times F_G(i,j)\}}{\sum_{i=1+U}^{M-U} \sum_{j=1+V}^{N-V} F_G(i,j)} \tag{9.7}$$

$d_M(F_G, F_P)$ is also directional, and the final distance is the smaller one out of $d_M(F_G, F_P)$ and $d_M(F_P, F_G)$. The distance definition is a simplified form of Formula 9.5 based on single bit coding. Only, the feature points coded with 1 are used for distance calculation instead of the whole pixels in image. So the denominator in Formula 9.7 presents the number of feature points in image G accordingly, and we denoted it by N_P. Such handling of distance calculation aims to reducing the negative effect of noise which may be preserved after binarization process. Although the binary feature image confines the matching to a smaller but more distinguishable region, the introduction of threshold in binarization would bring into intra-class variation as the threshold is sensitive to the image contrast. Especially when the illumination intensity changes with the band of the incident light, the binary feature image of one sample may be only part of another one from the same dorsal hand. According to the improved distance calculation as shown in Formula 9.7, at least one out of $d_M(F_G, F_P)$ and $d_M(F_P, F_G)$ is very close to 0.

Note that extending matching templates with pixel-to-area scheme may not be able to deal with potential case of large translation and rotation problems very well; we draw support from an algorithm called Iterative Closest Point (ICP) (Besl and McKay 1992), a useful method for registering two images. The point-to-point relationship is established through finding the closest feature points. It is further adjusted to new positions step by step, while the sum of point-to-point square distance converges to the nearest local minimum value monotonically without destroying original topology structure.

9.4 Optimal Band Selection

9.4.1 Left–Right Comparison

The large multispectral database is established on the foundation of assumption that the hand vein patterns are unique for not only each identity but also each hand. In Ahmed's work (Badawi 2006), the probability of the left–right dorsal hands for the same person to be judged having similar vein pattern could show the distance between left–right matching and genuine matching which use left or right dorsal hands only. This method equals to setting null hypothesis H_0 to be that genuine matching is similar as left–right matching, and then, the optimal threshold for the two kinds of matching is taken as the sample statistics. All distances are assumed to fit normal distribution, and we use Z-test to calculate p value. If p value is less than predefined confidence level, we can reject the null hypothesis and the left and right dorsal hands from the same person with large difference can be treated as two independent samples.

On the other side, whether the left–right matching distance is similar as the distance of imposter matching is not validated. Another hypothesis about this point needs to be set up. The expected result is that the corresponding p value is remarkably large than predefined confidence level and the second hypothesis can be adopted. Here, we use MFRAT as the feature tool to do the tests. Using all five multispectral groups, the experiment involves 4220 ($C_5^2 \times 422$) genuine matches, 88,831 (C_{422}^2) imposter matches, and 1055 (5×211) left–right matches. Figure 9.8

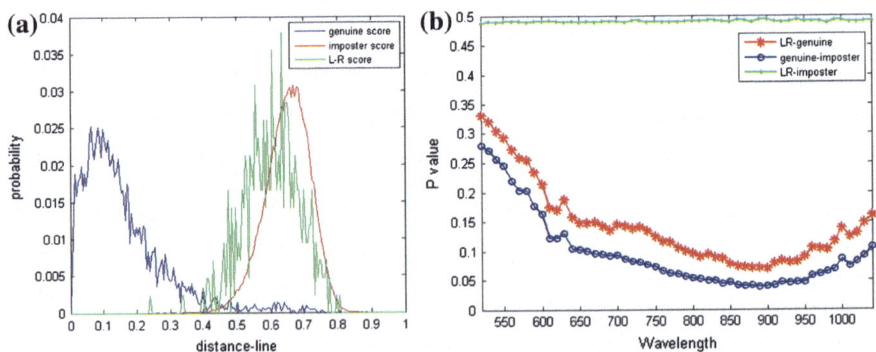

Fig. 9.8 **a** The probability density distribution of three kinds of matching. **b** The p values of LR-genuine (*red*) and LR-imposter (*green*) hypothesis based on all 53 bands. P value of original genuine–imposter matching (*blue*) is given out for comparison

gives out the probability distribution density of these three matchings and the p value of above two hypotheses.

The distribution of left–right dorsal hand matching is very close to imposter matching as illustrated in Fig. 9.8a. The red and green p value curves in Fig. 9.8b reflect LR-genuine distance and LR-imposter distance indirectly. The red curve can reach its minimum value about 0.07 on the band 900 nm; it is obviously that the distance between left–right matching and genuine matching is significant. However, the curve has rising trend on both sides, the direction of reducing and increasing wavelength from band 900 nm. This result reflects the difference identification of ability among difference parts of multispectral bands. Although the p value of red curve can be more than 0.15 on the bands below 650 nm, we can still reject the hypothesis because of the original poor performance on these bands. To illustrate this point, the p value of genuine and imposter matching (blue curve) is introduced. It is apparent that LR-genuine distance is very similar as genuine–imposter distance, though they both behave not well on visible part. On the other side, the p value of the second hypothesis, shown as the green curve, is very close to its theoretical maximum value 0.5 overall the bands, and it is concluded that the vein difference between left and right hands can be considered as difference from two independent samples. So the whole multispectral database can be expanded to a large database with 422 independent subjects.

9.4.2 Feature Comparison Result

In subspace learning method $(2D)^2$PCA, the proportion of cumulative energy is set to be 90 %. A template with 6 directions is chosen to obtain direction map in Gabor feature extraction, and the parameters are optimized as $\omega = 0.70, \sigma = 1.26, \gamma = 4$. MFRAT adopts 12-direction scheme to form the energy feature image. Apart from $(2D)^2$PCA, which focuses on training an image projection model, the other three methods can obtain feature maps as intuitive intermediate result as shown in Fig. 9.9.

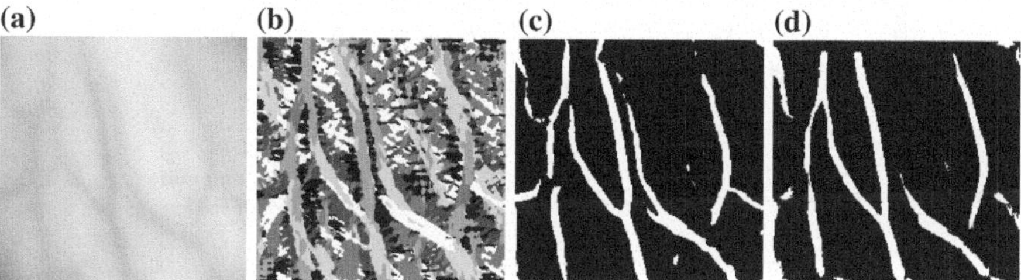

Fig. 9.9 The feature map based on different feature patterns. **a** Original image. **b** Direction map of real part based on Gabor filter. **c** Binary energy image based on Gaussian filter. **d** Binary energy image based on line extraction

The feature matching is operated on each single band separately. Leave-one-out cross-validation is adopted to get the first rank recognition result. As each subject has five multispectral groups, overall 712,336 (4×422^2) matches are calculated for each group as the testing set. The recognition scheme is conducted homogeneously on the three different feature patterns. The recognition result based on 53 bands is given out in the Fig. 9.10. As coding-based method uses the spatial distance measure like line-based method, the problem of alignment error in ROI image should be also resolved. We apply the similarity transformation matrix of ICP in line-based method on ROI image for Gabor feature extraction directly, though the process would get more time-consuming. The code is not optimized for efficiency, and here, we pay more attention to performance.

All feature patterns have the common properties that visible spectral part performs worse than NIR light. The two curves of CompCode and MFRAT both show that recognition declines down rapidly in the visible light part, especially when the band is lower than 600 nm. Because dorsal hand vein cannot get clear and high-contrast vein image under visible light, the lack of effective distinguishable texture would lead to unexpected bigger intra-class distance. On the other hand, only $(2D)^2$PCA recognition curve somewhat presents two peaks around the band of 580 and 880 nm, though recognition accuracy of 580 nm still cannot compete with it under 880 nm. We can conclude that surface skin texture is not robust or pervasive for a population with a large amount, especially in the aspect of fine texture expression. Obviously, NIR spectra focus on the vein feature which makes a good claim for the importance in identification.

If we only focus on NIR light part and settle the feature patterns with the descending order of recognition rate, they are, in order, MFAT, CompCode, and

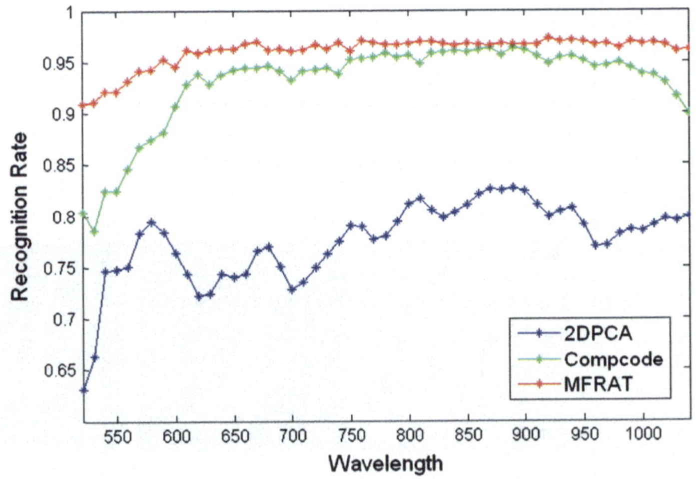

Fig. 9.10 The recognition results across the bands from 520 to 1040 nm based on three different feature patterns, $(2D)^2$PCA, CompCode, and MFRAT

$(2D)^2$PCA. Line-based method shows its superiority to other two feature patterns, because binarization process is exclusive to vein extraction, and the advantage is that the non-vein region of dorsal hand is not used for distance calculation. Coding the region without remarkable texture is not helpful to recognition rate improvement. Not only line-based method but also coding-based method has local adaptive capability because the feature extraction relies on the relationship between current pixel and its surrounding area. So they are both insensitive to illumination change across the whole spectra. $(2D)^2$PCA dose not have such good property, and it comes natural to have significantly lower recognition rate.

9.4.3 Feature Estimation

The highest recognition accuracy out of all 53 bands is taken as the basis to optimal feature selection. This measure is not enough to reflect the robustness of relevant feature pattern. It is assumed that the texture has a process of continuous change when the spectral wavelength increases uniformly, and the recognition rate should at least follow a relatively smooth curve. However, the recognition rate curves in Fig. 9.10 have unsteady fluctuation. Commonly, the number of limitation of testing samples would result in such phenomenon, and it can only be relived through setting up a large database. Here, slight instability of illumination intensity is another reason, and it would be finally reflected in the form of image contrast change. It is hard to guarantee that the light source intensity and system transmission are both constant in such a large spectral region, so a feature pattern robust to illumination change is important. Curve smoothness is taken as the criterion for features' robustness.

Let R be the recognition curve across all bands, and it is a vector with 53 dimensions. $\hat{R}$ is the curve fitting result of R through a Gaussian window with size of 7. The smoothness coefficient is the Pearson relationship between the two vectors. The closer the value is to 1, the higher degree of smoothness does the corresponding curve have. Table 9.1 lists the two estimation criterion values of three feature patterns. CompCode method with Gabor feature has the highest stability when bands change, and the corresponding smooth coefficient can get up to 0.9775. The smooth coefficient of MFRAT is not inferior much to the best robust feature, so it is still the first choice when only one kind of feature pattern is used.

Table 9.1 Estimation criterion values of single feature

Feature	Highest accuracy (%)	Smoothness coefficient	Optimal band (nm)
$(2D)^2$PCA	83.78	0.7168	870
CompCode	96.45	0.9775	870
MFRAT	97.27	0.9585	920

9.4.4 Feature Fusion

Considering that MFRAT has high identification ability and Gabor feature has high stability along with the band change, taking advantage of them by proper fusion scheme is proposed to pursue higher accuracy. Now that Gabor feature has advantage in describing fine texture, and the drawbacks mainly come from the low-contrast non-vein region, and we can combine the Gabor feature together with the vein binary image got from MFRAT method. It means that the Gabor feature is just limited within the image part with high line energy. Meanwhile, CompCode is reserved for final distance measure. The task can be implemented by feature-level fusion.

Figure 9.11 is an illustration of feature fusion image generation. If the two compared images come from different subjects, the lapped part may take very little account of the whole foreground region, so merging rule is taken to construct line feature union image and CompCode image from different samples which are added to the union region. The next matching scheme is the same as the coding-based method. The experiment shows that the maximum recognition rate can rise up to 99.21 %.

The great performance improvement owes to the reason that we just select the image region with the most robust feature for distance calculation. This fusion scheme can reduce intra-class distance remarkably, while the reduction of between-class distance is not great. If the two compared dorsal hand image comes from the same subject, the overlapping part accounts for a very large proportion of the union image. On the contrary, the combination of different vein shapes would introduce large non-overlapping region, in which the coded direction often appears

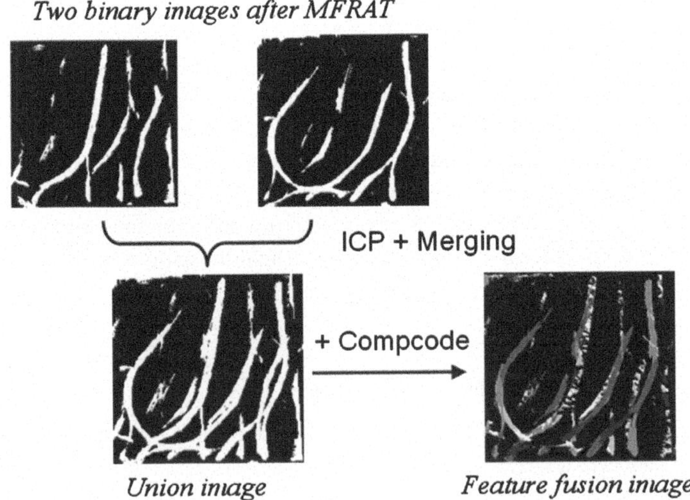

Fig. 9.11 An illustration of feature fusion image generation based on MFRAT and Gabor feature

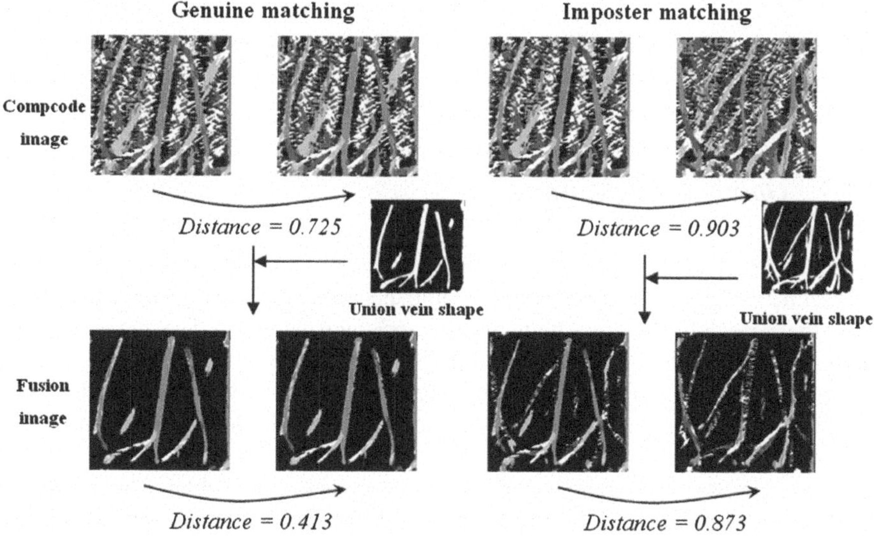

Fig. 9.12 The distance change after feature fusion of MFRAT and CompCode. The *left* column is about genuine matching with large distance decrease; the *right* column is about imposter matching with little distance decrease

to be messy, so between-class can not decrease obviously. It can be seen that this coding scheme is closely relevant to vein shapes. Figure 9.12 is two illustrations of distance comparison before and after feature fusion based on genuine and imposter matching.

Another fusion scheme is realized on the matching level. It needs normalizing different distance measure and obtains weighted sum as new distance of multiple features. If the fusion is only based on only two features, we can define a coefficient k to determine the weights as shown in Formula 9.8.

$$\begin{cases} k = \frac{\mathrm{EER}_2}{\mathrm{EER}_1} \\ d_{\text{fusion}} = w_1 d_1 + w_2 d_2 = \frac{k}{1+k} d_1 + \frac{1}{1+k} d_2 \end{cases} \quad (9.8)$$

EER_1 and EER_2 are equal error rate (EER) of corresponding two features. If $k = 1$, the formula is simplified to be non-weighted model. The estimation criterion values of fusion are listed in Table 9.2. The recognition accuracy of weighted

Table 9.2 Estimation criterion values of feature fusion

Fusion scheme		Highest accuracy (%)	Smoothness coefficient
Feature level (MFRAT + CompCode)		98.71	0.9713
Matching level (MFRAT + CompCode)	Non-weighted	96.45	0.9660
	Weighted	99.17	0.9800

MFRAT–Gabor fusion based on matching level has higher value (99.17 %) than feature-level fusion. In addition, the optimal feature fusion has the highest smoothness coefficient 0.9757.

9.4.5 Optimal Single Band

Weighted MFRAT–CompCode fusion method based on matching level is currently the best feature pattern combination for multispectral dorsal hand recognition. It can meet the demands for both high performance and stability. The corresponding optimal single band is 890 nm with recognition accuracy 99.17 %. Figure 9.13 is the recognition rate curve of this feature pattern. The top five bands are, by rank, 890, 880, 900, 870, and 860 nm. It can be found that the high performance concentrates on a narrow band region, which is centered by 890 nm. Generally speaking, spectra with larger wavelength can reflect more subtle vein on dorsal hand, so 890 nm wins the first place in presenting the richest information. If the spectrum wavelength gets small, the reflected images focus on the main vein network with large width, and the main vein network takes into account the majority of contribution to recognition. That is why the recognition rate declines slowly in the left part. On the other side, the subtle vein sometimes is not clear enough under bands more than 900 nm, and this is due to the testing samples' individual varieties such as fatness. Though the vein information is more abundant, these unreliable features are not suitable for high-performance system. So the curve falls down quickly in the band region above 890 nm. Viewing the results in a larger spectral scale, the bands 800–930 nm take up all top 14 places in recognition rate descending order. This makes certain again the important role of main vein network in dorsal hand recognition.

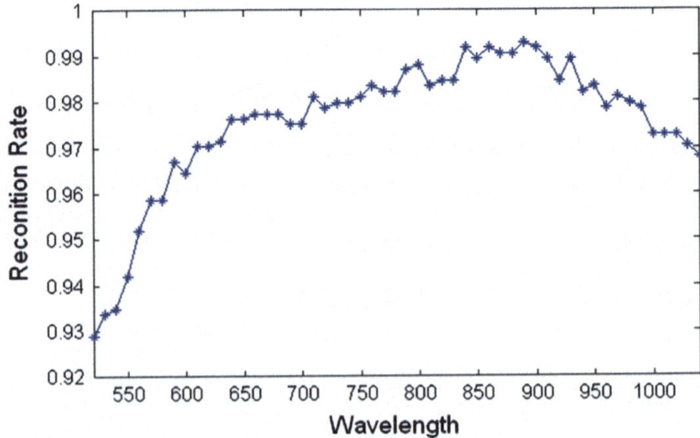

Fig. 9.13 The recognition rate curve of MFRAT–CompCode fusion from band 520 to 1040 nm

9.5 Summary

With the purpose of investigating the performance difference of various bands, we use multispectral technique instead of traditional near-infrared (NIR) for dorsal hand recognition in this work. Part of visible lights is also taken into account, and the capture system is set up with multispectral imaging. The fusion of modified infinite Radon transform and Gabor-based CompCode provides a convenient way for feature extraction with high recognition rate and preserving stability to face the illumination change across the whole spectra. The spectrum 890 nm is proven to be the best choice for dorsal hand recognition. This study provides a good guidance of band selection for not only actual applications, but also the research work on multispectral image in the future.

References

Badawi AM (2006) Hand vein biometric verification prototype: a testing performance and patterns similarity. IEEE Proc Image Process Comput Vis Pattern Recognit 2:26–29

Besl PJ, McKay ND (1992) A method for registration of 3-D shapes. IEEE Trans Pattern Anal Mach Intell 14(2):239–256. doi:10.1109/34.121791

Chen L, Zheng H, Li L, Xie P, Liu S (2007) Near-infrared dorsal hand vein image segmentation by local thresholding using grayscale morphology. In: 1st international conference on bioinformatics and biomedical engineering, pp 868–871. doi:10.1109/ICBBE.2007.226

Daugman JG (1993) High confidence visual recognition of persons by a test of statistical independence. IEEE Trans Pattern Anal Mach Intell 15(11):1148–1161. doi:10.1109/34.244676

Ding Y, Zhuang D, Wang K (2005) A study of hand vein recognition method. IEEE Proc Int Conf Mechatron Autom 4:2106–2110. doi:10.1109/ICMA.2005.1626888

Im S-K, Park H-M, Kim Y-W, Han S-C, Kim S-W, Kang C-H (2001) A biometric identification system by extracting hand vein patterns. Korean J Phys Soc 38(3):268–272

Jia W, Huang D-S, Zhang D (2008) Palmprint verification based on robust line orientation code. Pattern Recognit 41(5):1504–1513. doi:10.1016/j.patcog.2007.10.011

Kong A, Zhang D (2004) Competitive coding scheme for palmprint verification. Int Conf Pattern Recognit 1:520–523. doi:10.1109/ICPR.2004.1334184

Kumar A, Prathyusha KV (2009) Personal authentication using hand vein triangulation and knuckle shape. IEEE Trans Image Process 18(9):2127–2136. doi:10.1109/TIP.2009.2023153

Li W, Zhang L, Zhang D, Lu G, Yan J (2010a) Efficient joint 2D and 3D palmprint matching with alignment refinement. IEEE Conf Comput Vis Pattern Recognit 795–801. doi:10.1109/CVPR.2010.5540134

Li X, Liu X, Liu Z (2010b) A dorsal hand vein pattern recognition algorithm. In: 3rd international congress on image and signal processing, vol 4, 1723–1726. doi:10.1109/CISP.2010.5647776

Liu X, Zou B, Sun J (2004) A new approach to separating touching spots in particle images. In: Proceedings of international symposium on intelligent multimedia, video and speech processing, pp 133–136. doi:10.1109/ISIMP.2004.1434018

Matcher SJ, Elwell CE, Cooper CE, Cope M, Delpy DT (1995) Performance comparison of several published tissue near-infrared spectroscopy algorithms. Anal Biochem 227(1):54–68. doi:10.1006/abio.1995.1252

Ostu N (1979) A threshold selection method from gray-level histogram. IEEE Trans Syst Man Cybern 9(1):62–66. doi:10.1109/TSMC.1979.4310076

Ramalho SM, Correia P, Soares L (2011) Biometric identification through palm and dorsal hand vein patterns. IEEE Int Conf Comput Tool (EUROCON) 1–4. doi:10.1109/EUROCON.2011.5929297

Wang L, Leedham G, Cho S-Y (2007) Infrared imaging of hand vein patterns for biometric purposes. IET Comput Vis 1(3–4):113–122. doi:10.1049/iet-cvi:20070009

Wang L, Leedham G, Cho DS-Y (2008a) Minutiae feature analysis for infrared hand vein pattern biometrics. Pattern Recognit 41(3):920–929. doi:10.1016/j.patcog.2007.07.012

Wang K, Zhang Y, Yuan Z, Zhuang D (2008b) Hand vein recognition based on multi supplemental features of multi-classifier fusion decision. IEEE Proc Int Conf Mechatron Autom 1790–1795. doi:10.1109/ICMA.2006.257486

Wu X, Gao E, Tang Y, Wang K (2010) A novel biometric system based on hand vein. In: 5th international conference on frontier of computer science and technology, pp 522–526. doi:10.1109/FCST.2010.65

Yang J, Zhang D, Frangi AF, Yang Y-J (2004) Two-dimensional PCA: a new approach to appearance-based face representation and recognition. IEEE Trans Pattern Anal Mach Intell 26(1):131–137. doi:10.1109/TPAMI.2004.1261097

Yuan W, Wang R, Sun S (2010) Palm-dorsa vein recognition based on two-dimensional fisher linear discriminant. Chin J Comput Appl 30(3):646–649

Yuksel A, Akarun L, Sankur B (2010) Biometric identification through hand vein patterns. In: International workshop on emerging techniques and challenges for hand-based biometrics, pp 1–6. doi:10.1109/ETCHB.2010.5559295

Zhang D, Kong W-K, You J, Wong M (2003) Online palmprint identification. IEEE Trans Pattern Anal Mach Intell 25(9):1041–1050. doi:10.1109/TPAMI.2003.1227981

Zuo W, Zhang D, Wang K (2006) Bidirectional PCA with assembled matrix distance metric for image recognition. IEEE Trans Syst Man Cybern B Cybern 36(4):863–872. doi:10.1109/TSMCB.2006.872274

Chapter 10
Multiple Band Selection of Multispectral Dorsal Hand

Abstract The optimal single band for dorsal hand recognition is 890 nm based on the fusion of MFRAT and CompCode. The extracted vein information is limited within one spectrum only. One of the advantages of multispectral technique is to pursue higher recognition performance through the fusion of multiple bands. The main theoretical basis is that the images on different bands have complementary information, which is helpful to performance improvement. Obviously, adding more bands is similar to feature dimension increase, and the redundant information may also increase quickly especially when the bands are highly relevant to each other. So the multispetral band selection process is limited to high efficiency, and it demands that the selected bands have the maximum uncorrelation. Noting that recognition of visible region is obviously far worse and the texture of non-vein part is not reliable enough for efficient recognition, we plan to do the multiple band selection work in near infrared (NIR) region only. 422 dorsal hands of East Asians with different bands ranging from 700 to 1040 nm are used. This chapter tries to address two basic issues, the number of the bands for optimal group and which bands can explain the multispectral model more precisely. Unlike optimal single band selection, exhaustive method is not practicable for this task. The number of possible combination is immeasurable, especially when the optimal band number is unknown. Our scheme is to realize the task in two steps: First, divide the NIR region into several band classes according to special rules so that the number of optimal band is fixed; secondly, choose the bands from these classes to represent them with proper estimation criterion.

Keywords Dorsal hand recognition · Multispectral image · Band clustering · Band fusion

10.1 Introduction

Dorsal hand recognition, as a very important aspect in biometric family, always performs well for its superior characteristics of uniqueness and permanence during one's whole life (Wang et al. 2008). The study of dorsal hand has achieved

D. Zhang et al., *Multispectral Biometrics*,
DOI 10.1007/978-3-319-22485-5_10

substantial improvement in the last decades. Researchers made great effort to obtain higher recognition performance in different aspects, such as imaging ways (Assaleh 2011), more precise vein extraction (Ding et al. 2005), and feature representation ways (Wang et al. 2007). Even the influence from different external conditions such as pressure and temperature is also taken into account (Yuksel et al. 2010). Besides, the systems for dorsal hand recognition are widely published (Wu et al. 2010; Sanchit et al. 2011). More and more fusion systems based on multiple modalities spring up in recent years. According to the requirement for convenience in image collection process, the fusion objects to dorsal hand are often hand-based features such as fingerprint, palmprint, and hand shape.

Hand vein is sensitive to the light source change. The reflected vein change is not only limited to the gray change of captured image, but also related to the vein shapes, width, integrity, and so on. The texture difference makes it possible to analyze performance improvement after band fusion, and it is a good trail of thought to capture dorsal hand image with more information for complement. Multispectral imaging is introduced in dorsal hand recognition. The database could be seen as a 3-D data if the band group is linked consecutively to form a new dimension. Unlike multimodality biometrical recognition, band fusion is classified into multisensor biometrical recognition, because the reflected images come from the same source. As the bands for study are continuous and the wavelength of neighboring bands does not differ very much to the others, redundant information takes up the majority part in these high dimensional datasets. A traditional way of fusion is to connect all the features in series which are usually transformed to a normalized form, and this approach often results in a fusion feature with large dimension, which may further reduce recognition efficiency largely. In practical view, a multispectral biometrical recognition system is expected to have high performance with only a few light source kinds. So a basic rule of band fusion is to choose as few bands as possible to represent the whole spectra region, and band selection plays an important role before band fusion.

Multispectral technique is proposed to enhance image in its earliest use (Chen et al. 2010). Once it is applied to recognition domain, the investigation of complementary and redundant information for band selection comes up to an important issue. This point can be reflected in various applications, not limited to biometrics only. Huang and Li (2008) used search-based method to realize band selection in remote sensing. It sorts the bands relying on various criteria and uses different sequential search algorithms, such as sequential forward selection and sequential backward selection, to form the best optimal combination. Arnau et al. (2004) used information-based methods to extract optimal subsets of spectral images in food safety detection. The minimization of dependent information criterion is proposed based on the condition that the conditional entropy is maximized. Ekenel et al. (2010) applied traditional recognition-based methods for multispectral face recognition. In his experiment, a sliding window with varying size is used to analyze the relation between bands number and performance.

Theoretical study about multispectral band selection can be classified to feature selection domain if each band is considered as feature. They both have function of reducing dimension for shorter computing and better generalization. Applying mature technique from feature selection directly on band selection is a good way. Feature selection model consists of four modules (Fig. 10.1); they are generation, evaluation, stopping criterion, and validation module. Researchers usually control the generation and evaluation modules to construct specific feature selection model. Generation module is used to generate subsets of feature for evaluation with various schemes, such as complete scheme, heuristic scheme, and random scheme. Complete scheme generates exhaustive possible subsets following the rules of breadth first or branch first. This exhaustive method would no doubt present a challenge for storage and computing. The widely used scheme is heuristic scheme, which contains sequential forward selection, backward selection, and other variants of these methods, such as bidirectional search. Subset generation of random scheme relies on some algorithms such as simulated annealing and genetic algorithms. Obviously, the latter two schemes can generate subset of features more quickly than complete scheme and achieve a quite ideal result. However, being trapped in local minimal point is an ineluctable problem, so reducing the hidden risk while keeping the high efficiency of band selection is the main task of this chapter.

According to the ways of evaluation, this module can also be classified into 3 kinds. Wrapper way uses a predictive model to score feature subsets, and it is the direct way. Filter way uses a proxy measure instead of the error rate to score a feature subset. Fox example, the original data are usually transformed to mutual information or correlation coefficient to reflect the relationship of features among the chosen subset. The trick of filter way is that it can promote efficiency greatly by replacing evaluation target. The last one is called embedded way, and it performs feature selection by means of another model construction. Nevertheless, embedded ways are often limited to the subset searching scheme of the model itself, so it usually combines with specific generation scheme and it cannot be used flexibly. Take the commonly method Lasso for instance, the model searches better subset step by step, and it would fall in local minimal point because it has to use heuristic scheme to generate feature subsets. It is obvious that taking use of filter way and searching for a convenient and rapid evaluation function to address the problem of high feature dimension is also important for the task of band selection.

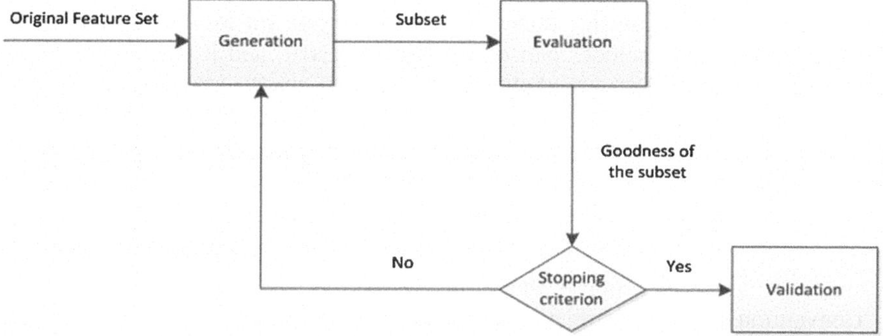

Fig. 10.1 Four modules of feature selection model

To summarize, the subset generation module requires high efficient method, while the evaluation modules need easy calculated evaluation function which is equivalent to error rate. Considering that the final selected bands should have the least proportion of redundancy to overall information, correlation is worth studying to reveal the independence of these bands. Meanwhile, it is assumed that the final combination of selected bands should have least redundancy, so it can be seem as the center band of several band clusters. Our scheme is that band selection can proceed in two phases, band clustering and choosing representative band. An unsupervised clustering method is firstly proposed to solve the problem rapidly: Which band clusters is the best division? In fact, we need to choose the proper number of clusters. Keeping more bands for selected combination although can reserve the majority of distinguishable information to ensure high recognition rate, the remained calculation complexity would make band selection meaningless. This is a trade-off between the total amount of texture information and band number. So a good iterative stopping condition should be designed for unsupervised clustering method.

10.2 Correlation Measure

The purpose of our research is to seek for the optimal band combination to present the whole spectral range. These bands should not only be representative for each spectral subsection, but also have the largest uncorrelation for distinguishing them. The uncorrelation, as a distance measure, should be robust enough to describe the texture difference directly.

10.2.1 Feature Representation

The multispectral acquisition system collects a total of 422 different hands including both left hands and right hands. Strictly speaking, near infrared (NIR) short wave light starts from 780 to 1100 nm. As the absorption spectrum of deoxygenated hemoglobin converges on 700–900 nm and gets its peak points on 780 nm, we spread the spectral region for dorsal hand capturing down to 700 nm and up to 1040 nm. This range includes part of red light and NIR, and it is wide enough to exhibit the texture change remarkably for further effective band clustering.

As the previous chapter noted, up to 2110 images from 422 East Asian volunteers are collected to construct multispectral dorsal hand database and 10 nm is chosen as the interval of the bands (Chen et al. 2011). All images are gray color with resolution of 501 $\times$ 512 pixels. We denote multispectral image by X_{ij}^{b}, which means the image from *j*th multispectral group on *b*th band of *i*th individual. A multispectral image group is shown in Fig. 10.2.

Conventional image correlation calculation is not stable enough for distance measure because of non-uniform illumination. A new kind of variable for single

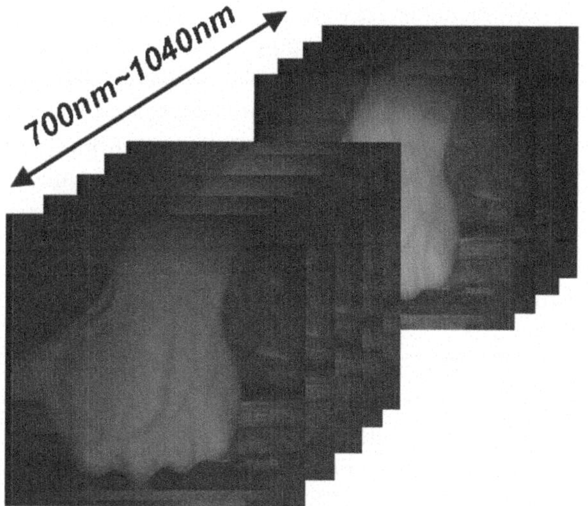

Fig. 10.2 A group of dorsal hand multispectral images. 53 bands ranging from 520 to 1040 nm are placed in order from short wavelength to long wavelength

band's characteristics representation should be found. Image preprocessing and robust feature extraction are necessary to build the correlation calculation on the foundation of feature level. After image mean and variance normalization, the region of interest with size of 224 * 224 is extracted to get the main part of dorsal hand, so position and rotation normalization is achieved. Radon transform can mark out the line features of hand vein with energy which reflect the gradient information in a limit local region (Jia et al. 2008). This vein extraction method is of very benefit to multispectral images, as it mainly extracts the principal veins under dermis and it is not sensitive to other textures such as noise-like skin surface and fine hair which may be imaged more clearly under short wavelength light. Compared with structure extraction method like the operations of vein thinning, pruning (Yang et al. 2010; Kumar et al. 2009), Radon transform can avoid information loss so that the slight difference between neighboring bands can be reserved. In addition, the light intensity change across the whole band range would not impact greatly on feature extraction because of its local area method. Figure 10.3 illustrates two results of feature extraction based on 780 and 880 nm. The latter one has sharper and smoother boundary of hand vein because of better absorption rate and less influence of fine hair on the back of hand.

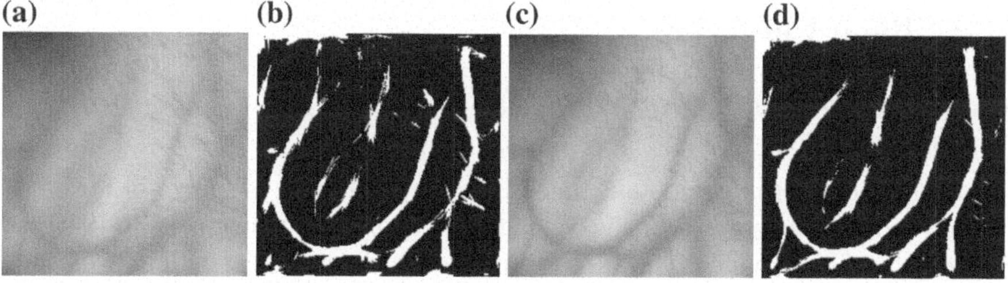

Fig. 10.3 ROI images after preprocessing and Radon transform results. **a–b** 750 nm; **c–d** 880 nm

All feature images are normalized to have the binary form, and each one is divided into non-overlapping blocks of 16×16 pixels size in our proposed approach. Afterward, the number of feature points in each block is counted, and they are concatenated to construct the overall feature vector F_{ij}^{b}, which has a length of 196. Each element has the value range from 0 to 255. Then, the feature vectors are all normalized with proper scaling to have the same sum of squares as 1.

10.2.2 Pearson Correlation

Pearson (1896) correlation coefficient between two variables is defined as the covariance of the two variables divided by the product of their standard deviations. It mainly reflects the two variables' linear relationship. When it is more close to 1, the changes in two compared variables converge exactly to the same trend.

$$r_{X,Y} = \frac{\mathrm{cov}(X,Y)}{\sigma_X \sigma_Y} = \frac{\sum_{i=1}^{n} (X_i - \overline{X})(Y_i - \overline{Y})}{\sqrt{\sum_{i=1}^{n} (X_i - \overline{X})^2} \sqrt{\sum_{i=1}^{n} (Y_i - \overline{Y})^2}} \tag{10.1}$$

All 5 groups of multispectral images are used in correlation calculation. So each band has a submatrix with 2110 feature vectors. We denote it by BM_b as shown in Formula 10.2. Here, N is 422 according to the number of subjects.

$$\mathrm{BM}_b = \left[F_{11}^{b}, F_{21}^{b}, \ldots, F_{N1}^{b}, F_{12}^{b}, \ldots F_{N2}^{b}, \ldots, F_{15}^{b}, \ldots, F_{N5}^{b}\right] \tag{10.2}$$

Only the feature images of the same subject from different bands with the same group index are chosen to be used to obtain the correlation coefficients; otherwise, the position deviation would give out unauthentic values. The correlation of any two different bands is the mean result of 2110 comparisons and considered as an element in the final correlation matrix RM.

$$r_{b_1 b_2} = \mathrm{corr}(\mathrm{BM}_{b_1}, \mathrm{BM}_{b_2}) = \frac{1}{5N} \sum_{i=1}^{N} \sum_{j=1}^{5} \mathrm{corr}\left(F_{ij}^{b_1}, F_{ij}^{b_2}\right) \tag{10.3}$$

10.3 Band Clustering

10.3.1 Correlation Map Analysis

The correlation matrix can be illustrated by correlation map as shown in Fig. 10.4. The x-axis and y-axis present the wavelength of bands ranging from 700 to 1040 nm. The gray level in each block shows the correlation between corresponding two bands.

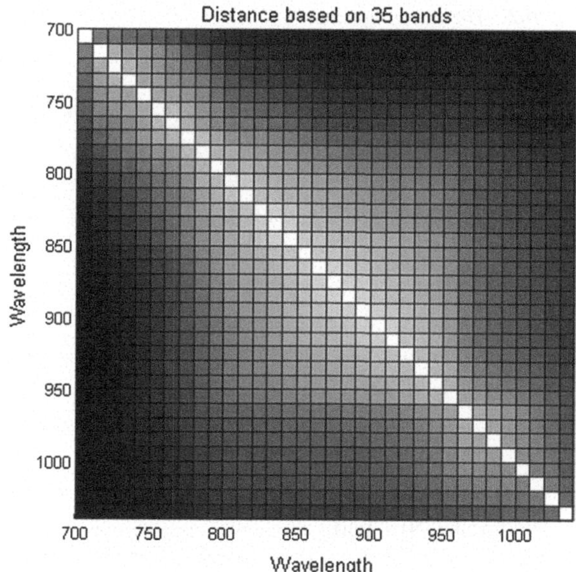

Fig. 10.4 Correlation map based on 35 bands

The darker block means the smaller correlation, that is to say the two related bands have large distance.

We can find 2 issues of correlation change among bands by observation of correlation map obviously.

1. Generally speaking, the change of dorsal hand's texture is considered to be consecutive with the light source wavelength increasing. So we held the conception that dorsal hand's reflectivity has continuous change on different bands. This assumption is proven by the correlation map that the blocks closed to diagonal line have comparatively higher gray values. However, the texture change rate is not a constant. If we limit the neighboring region within a narrow area like 2 bands and construct a 5×5 mean template, move it along the diagonal line to show the changing trend of mean correlation from certain band to other bands in its neighborhood. We call it local correlation analysis. It reflects the degree of dispersion of neighboring bands in a small region. Figure 10.5 plots the local mean correlation change based on different center bands.

   The correlation change is smoothed with curve fitting. We can conclude the changing trend into three stages. The first is the gentle rising stage corresponding to the red light part from 700 to 780 nm. NIR part below 930 nm is stable stage, which just has the highest performance in dorsal hand vein recognition according to the experimental results in previous chapter. The last stage reveals a sharp decline of local correlation value. It implies that the bands with larger wavelength in NIR part (930–1040 nm) have larger dissimilarity to the

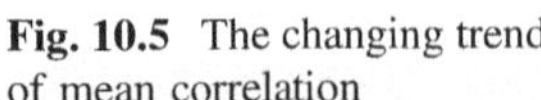

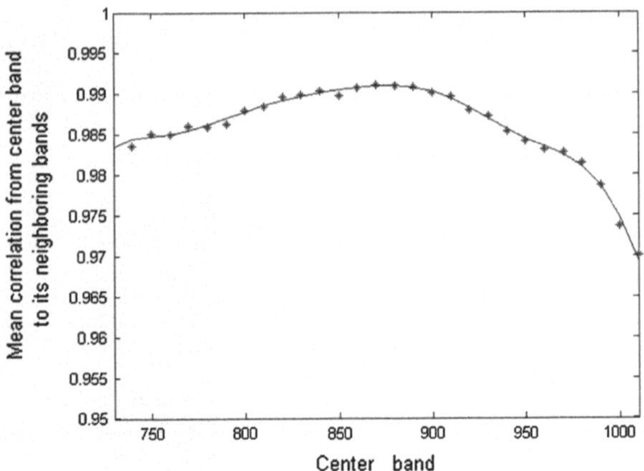

Fig. 10.5 The changing trend of mean correlation

neighboring bands compared with bands of in former two parts, and we call it incompact NIR part.

2. Each row or each column in correlation map represents a vector of correlation from a certain band to all other bands. The global correlation analysis describes the bands' distribution in a large scale. We choose 3 bands 720, 850, and 1000 nm to explore the global correlation change based on 3 typical parts; they are red light part, stable NIR part, and incompact NIR part divided by local correlation analysis.

A definition called spectrum distance is used to distinguish the band distance. It is the wavelength subtraction result of compared two bands. If the spectrum distance gets bigger, the correlation value decreases monotonously for all bands in NIR region. Take the band 720 nm for instance in Fig. 10.6a, and it has the lowest correlation value with the incompact NIR part. The low correlation values between different parts give a clue for the fusion of them with the purpose of recognition improvement.

According to the conclusion of local correlation analysis, the incompact NIR part should have more discrete distribution than red light part. Nevertheless, Fig. 10.6b shows that the distance from incompact NIR part to stable NIR part is not as significant as the one from red light part to stable NIR part. To explain the seemingly self-contradictory result, we set up a multispectral model for dorsal hand.

10.3.2 Model Setup

Each band can be seen as a point in a high dimensional space. The correlation value is just circular cosine of the angle formed by two related bands. If the two bands are further considered as two vectors, the intersection angle value reflects the distance between them. The form of distance expression can be transformed to arc length on

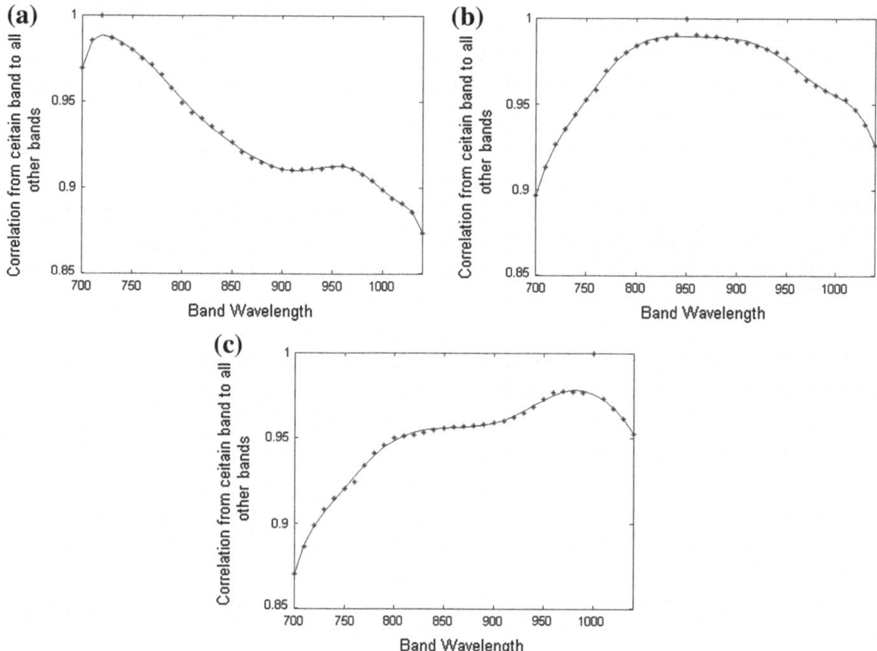

Fig. 10.6 Global correlation value based on 3 typical bands. **a** 720 nm; **b** 850 nm; **c** 1000 nm

the surface of sphere in high dimensional space when all bands are set to be unit vectors. As the radian value can replace the length directly, the distance matrix (dissimilarity matrix) can be calculated as the formula below.

$$\mathrm{DM} = \arccos\,(\mathrm{RM}) \tag{10.4}$$

The precondition of clustering work is to assume that the band change is basically a linear change. As the spectrum distance is only 10 nm between any two adjacent bands, the texture difference is so little that the bands change along the same direction in the context of small band scale. On the other hand, the slight direction change cannot be ignored when the change is accumulated to a certain extent.

Figure 10.7a is an illustration of 3 bands' distribution in 3-D. The unit vectors are labeled with $\overrightarrow{B_1}$, $\overrightarrow{B_2}$, and $\overrightarrow{B_3}$. The three vertex positions form a triangle $\Delta B_1B_2B_3$, whose edges can be regarded as the approximation of distance between bands. Setting the three bands to have consecutive wavelength in order, the angle $\angle B_1B_2B_3$ corresponds to the edge e_{13} with the maximum length, and it is generally an obtuse angle. So the edges e_{12} and e_{23} are basically on the same line when the triangle is projected onto a proper plane.

Extending the three bands model to the whole multispectral band group, 35 bands have an approximate projection result as shown in Fig. 10.7b. Vectors are

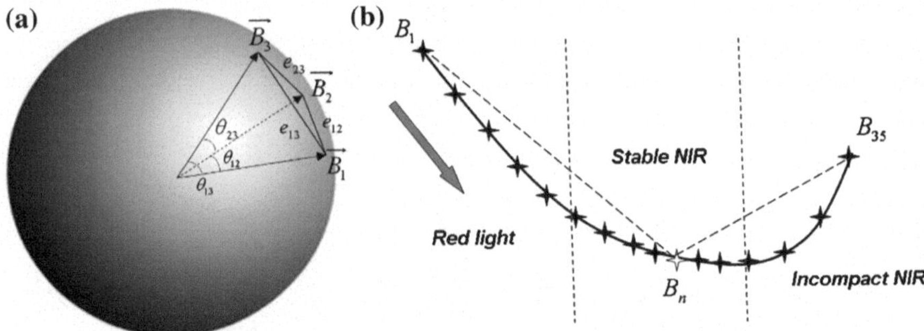

Fig. 10.7 The model setup for dorsal hand. **a** The schematic drawing of the three band vectors on sphere in 3-D space. **b** Approximate projection result of bands on 2-D plane

used to connect each two adjacent bands, and the connection of all vectors can form a directional curve from band B_1 to B_{35} on the plane of projection. The model of band distance is established on the basis of three principles.

1. The bands are not equally distributed in different parts.
2. Linear band change evolves into nonlinear when it expands to the whole spectral scale.
3. The directional curve appears to turn back in incompact NIR part.

10.3.3 Clustering Methodology

Searching for the combination of bands having the maximum uncorrelation is to resolve the question: Which clusters are the optimum clustering results? The consecutive texture change constrains that the bands in any final class should have wavelength continuity. Agglomerative hierarchical clustering technique is just the method which can proceed by a series of successive mergers from initial individual bands (Ozawa 1983; Johnson et al. 1998).

Given the same spectrum distance, visible light bands have comparatively larger distance than the ones in NIR region. The uneven data distribution indicates that it is not suitable to apply classic hierarchical clustering method directly to the distance model; otherwise, the clustering would be no doubt trapped in a trouble that several single visible bands with large distance to all other bands are clustered to be several classes individually. We introduce local principle to hierarchical clustering to solve the problem. It means that the band distance is only valid when it is used in a limit neighboring band range; otherwise, it should be normalized with local band distribution in global scale. The proposed method includes two steps: initial segmentation and improved hierarchical clustering.

The purpose of initial segmentation is to make sure that each segment includes at least 2 bands, which is necessary to normalize the band distance with local distribution. Initial segmentation aims to find out the large distance between two adjacent bands and extract the cutoff points to separate them. Local principle is also applied in this step. The cutoff point extraction is completed from a series of neighboring bands within limited region.

Assume there are N individual bands. In order to find out cutoff points, a template with 2-D Gaussian function is used. First, we predefine the minimum number of bands for each initial segment with Ls, which also determines the size of Gaussian template. The minimum number restricts the coverage of local principle. For example, if the minimum number $Ls = 2$, the two adjacent cutoff points should have a minimum spectrum distance of 2, and the template's size could be $2 * Ls = 4$. We denote the Gaussian template by $T(2 * Ls, \sigma)$ and add it to checkerboard kernel (Theodoridis 2006), which is composed of 1 and -1. The composite template H has a size of $2Ls \times 2Ls$ with the form as shown in the formula below. σ is the standard deviation of the Gaussian template.

$$H = \begin{bmatrix} -1 & \cdots & -1 & 1 & \cdots & 1 \\ \vdots & \ddots & \vdots & \vdots & \ddots & \vdots \\ -1 & \cdots & -1 & 1 & \cdots & 1 \\ 1 & \cdots & 1 & -1 & \cdots & -1 \\ \vdots & \ddots & \vdots & \vdots & \ddots & \vdots \\ 1 & \cdots & 1 & -1 & \cdots & -1 \end{bmatrix} * T(2 * Ls, \sigma) \tag{10.5}$$

The template H moves along the diagonal line of distance matrix to get a vector called cutoff value vector. If the center of template H matches a potential cutoff point, its left Ls bands and its right Ls bands should be divided into two separated segments. So the distances among bands within segments should be small, and we multiple -1 in checkerboard kernel along the diagonal line. On the other hand, the bands classified into different segments have large dissimilarity, and we multiple 1 in checkerboard kernel along the direction perpendicular to the diagonal line.

Not all local maximum points in cutoff value vector should be judged as cutoff points unless they match certain requirements. Taking each candidate cutoff points as center, we extract a vector V_c with a length of $2 * Ls - 1$. Because the spectrum distance between two neighboring cutoff points should not be shorter than Ls, the vector V_c cannot have more than one cutoff points. Then, the left and right parts of V_c are denoted by V_{left} and V_{right}, respectively, each of which has the length of $Ls - 1$. Once the center cutoff value is just the maximum one in vector $V_c(i)$, we could further use the rule in Formula 10.6 to decide whether the cutoff point candidate is adopted.

$$J_{\text{rate}} = \max(V_c)/\max(\min(V_{\text{left}}), \min(V_{\text{right}})) \tag{10.6}$$

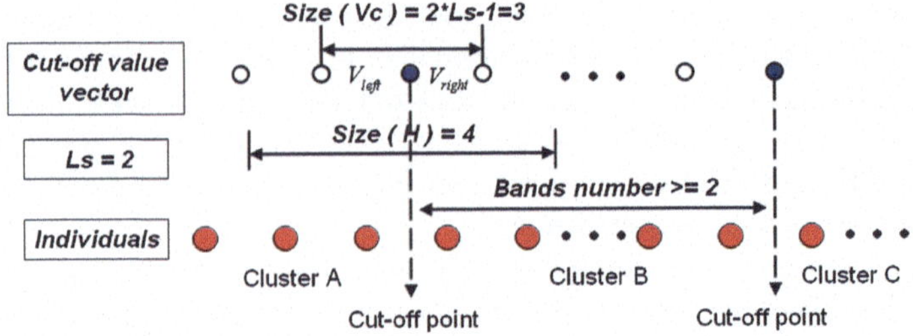

Fig. 10.8 Cutoff point extraction with local principle

Only if the element $J_{\text{rate}}(i)$ is not significantly smaller than mean(J_{rate}), the corresponding spectra would be preserved as cut-off point. Figure 10.8 is an illustration of cutoff point extraction.

Agglomerative clustering method is a "bottom-up" approach, which is useful for clustering raw data in order with iterative steps based on the data distribution. This method builds the hierarchy from initial segments to only one final cluster by progressively merging clusters and find out the merging order. In each iterative step, the two closest segments are assigned into a new segment according to the distance between two clusters, which is represented by distance of clusters' center bands or mean distance of all band pairs.

Conventional hierarchical clustering can adapt to the data clustering with continuous change and make sure spectra wavelength continuity in the same cluster. However, it is not competent to deal with the issue of uneven data. Set $S_1, S_2, S_3, \ldots, S_n$ to be the cluster centers after initial segmentation, and they also conform to the multispectral uneven distribution model. The segments in red light part have larger cluster distance, take S_1 for example, and it has the largest cluster distance to all other segments, even it may form a separate cluster itself unless the merging process fully completes. Figure 10.9b is the expected clustering result, as merging several segments with similar distribution seems to be more reasonable. This approach is similar to ratio-cut in graph theory.

We apply between-cluster distance as local information for normalizing within-cluster distance. The related distances are defined as the formula shown below.

$$S_B(C_i, C_j) = d\left(\overline{\text{BM}^i}, \overline{\text{BM}^j}\right), \quad i \neq j \tag{10.7}$$

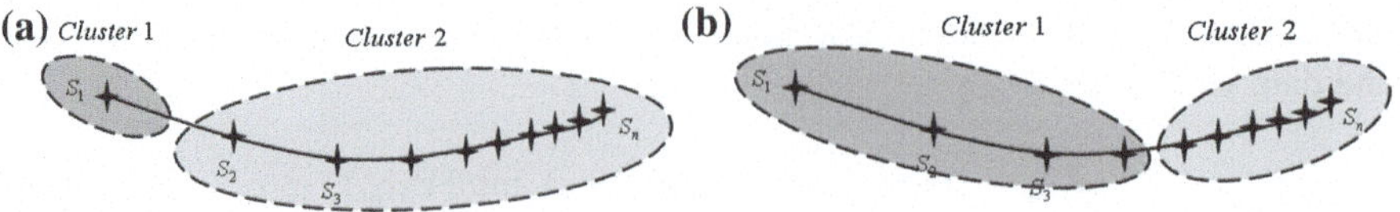

Fig. 10.9 An illustration of hierarchical clustering result with two clusters based on multispectral model. **a** Conventional hierarchical result. **b** Expected result

$$S_W(C_i) = \sum_{n=1}^{N_i} d(\overline{\mathrm{BM}^i}, \mathrm{BM}^i_b)/N_i \quad \mathrm{BM}^i_b \in C_i \tag{10.8}$$

Here, C_i and C_j are two clusters, BM^i_b means the band feature matrix of bth band in ith cluster. Between-cluster distance $S_B(C_i, C_j)$ can be calculated from the distance of two clusters' mean matrix $\overline{\mathrm{BM}^i}$ and $\overline{\mathrm{BM}^j}$, which are the mean value of multiple band matrix belonging to the same cluster. Set the initial segment number to be N_s, and the details of clustering process can follow these steps (Foote 2000):

1. Given two neighboring ith segment and $(i + 1)$th segment, calculate $S_B(C_i, C_{i+1})$, $S_W(C_i)$ and $S_W(C_{i+1})$.
2. Get $J(i) = S_B(C_i, C_{i+1})/(S_W(C_i) + S_W(C_{i+1}))$ and generate vector J.
3. Find out the minimum element $J_{\min}$ with its corresponding index j in J. Then merge jth segment and $(j + 1)$th segment into one segment. $N_s = N_s - 1$. Return to Step 1 unless $N_s = 1$.

10.3.4 Clustering Result

The cutoff point is presented by spectrum wavelength. For example, if a cutoff point exists between two neighboring bands 800 and 810 nm, we mark the cutoff point with 810 nm. The curve of cutoff value based on $Ls = 2$ is plotted in Fig. 10.10a. A total of 6 cutoff points are extracted to separate all 35 bands into 7 initial segments.

All initial segments can be merged into one cluster after 8 iterative steps of improved hierarchical clustering. In each step, one cutoff point with the smallest ratio of between-cluster to within-cluster distance should be deleted and the deleted sequence represents the merging order.

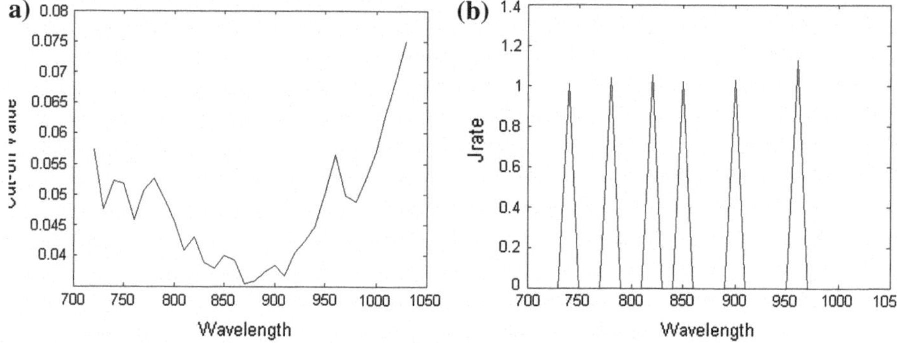

Fig. 10.10 **a** The curve of cutoff value. **b** J_{rate} of cutoff point candidates

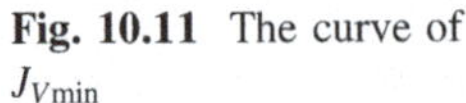

Fig. 10.11 The curve of $J_{V\min}$

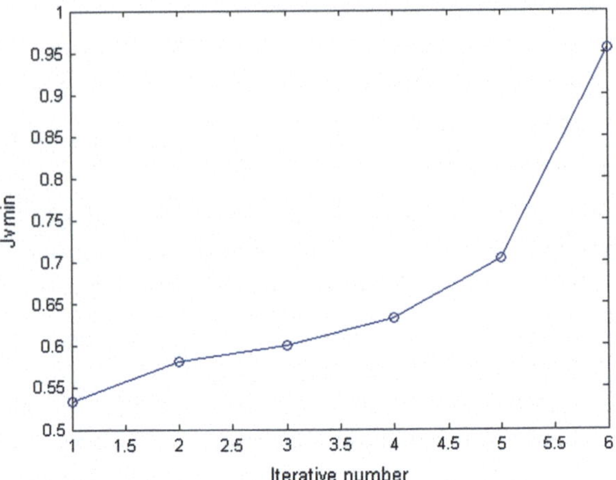

One of the most important tasks in band selection is finding a criterion for determining the number of final clusters. $J_{\min}$ got in each iterative step is competent to measure the merging necessity of corresponding two adjacent segments. We build a vector $J_{V\min}$ composed of all $J_{\min}$ value in the iterative order. $J_{V\min}$ increases generally as shown in Fig. 10.11. It means that the merging necessity decline, while individual class getting larger with more bands.

$J_{V\min}$ changes smoothly except the last two values. It indicates that the normalized distance between any two neighboring segments is so large that forcing them together is not appropriate. So we reserve the last two cutoff points 780 and 960 nm and divide the whole spectra range into 3 parts. Figure 10.12 displays the final results of clusters with the clustering order.

The dorsal hand vein information has large dissimilarity among three light source components, red light and two parts of NIR light. A large range of spectra from 780 to 950 nm maintaining to be the same class indicates that the difference under stable NIR light is not significant. So the complementary information exists mainly in other two parts.

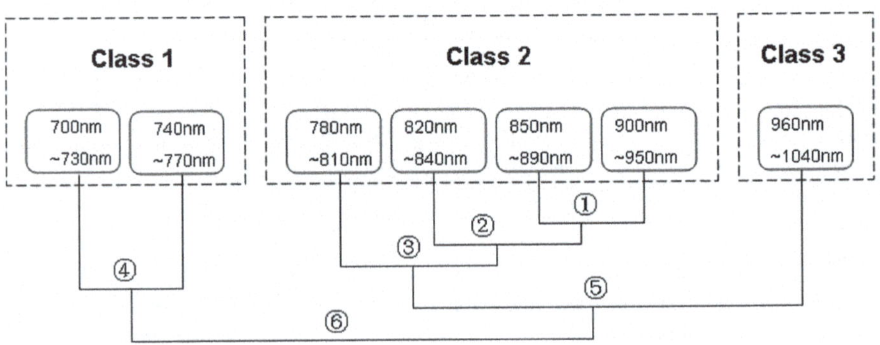

Fig. 10.12 Clustering order and result

Table 10.1 Robustness analysis of initial segmentation

No.	$Ls = 2$		$Ls = 3$		$Ls = 4$	
	Cutoff points (nm)	J_{min}	Cutoff points (nm)	J_{min}	Cutoff points (nm)	J_{min}
1	900	0.535	890	0.548	960	0.704
2	850	0.581	850	0.635	780	0.955
3	820	0.601	960	0.704		
4	740	0.632	780	0.955		
5	960	0.704				
6	780	0.955				

10.3.5 Parameter Analysis

In process of getting initial segments, a Gaussian template $T(2 * Ls, \sigma)$ is used and Ls determines the valid range of local principle. In this test, we set Ls to be other values such as 3 and 4. Meanwhile, we use new standard variance $\sigma' = \sigma * (Ls)/2$ for Gaussian template to adapt the template size. It is predicted that the larger value of Ls would result in smaller number of initial segments. The deleted order of cutoff points and their J_{min} values are listed in Table 10.1.

The test aims to investigate the robustness of the clustering result. Through changing the initial conditions, the last three reserved cutoff points are the same. It implies that the 3 clusters have high degree of stability and the uncorrelation among them are so large that all tests can converge to the same conclusion despites of the different iterative step number. The clustering result just coincides with the three square highlighted regions along the diagonal line of correlation map in Fig. 10.4.

10.4 Band Selection

10.4.1 Representative Band Selection

The applications of multispectral technique to dorsal hand are built on regarding the images of various bands as multiple sources and making use of complementary information for the purpose of recognition accuracy improvement. Now that 3 band clusters have information redundancy to the minimum extend, and it is natural that the bands of optimal combination come from these clusters. It means that selecting one band from each cluster to form band subset is enough for evaluation. The advantage of this scheme is that it can reduce the number of tested subsets greatly. Because exhausted method for subset generation is in the constraint of band clustering result, we can call it semi-exhausted scheme.

On the other hand, we need that the selected band in each cluster is highly representative. The bands on the edge of clusters may weaken information complementary ability of the band combination. So it is expected that the selected bands

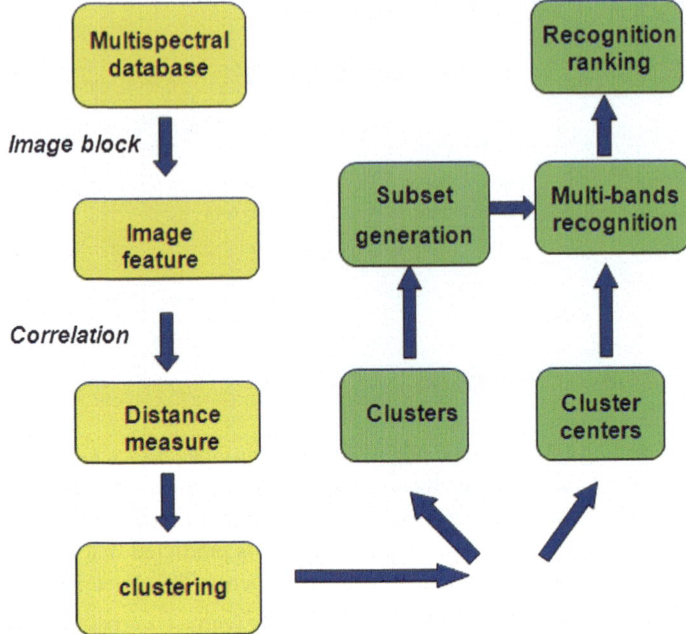

Fig. 10.13 The framework of band selection

are all close to the cluster center. According to this point, additional constraint is brought into the process of band fusion. On the foundation of matching-level fusion, we use the distance between selected band and cluster center method as penalty term to weight the distance of single band and further recalculate the fusion distance.

Combining described before, the whole framing of band selection is shown in Fig. 10.13. The left and right columns are about band clustering and band selection, respectively. The band combination which has the highest recognition rate is chosen as the optimal result.

10.4.2 Fusion Results

We take leave-one-out scheme in recognition ranking. Each group of multispectral images is chosen to be the training samples iteratively, and the others for testing. This process repeated five times. The best single feature Radon transformation is chosen to extract dorsal hand vein for recognition. According to the single band recognition result in Chap. 9, the highest rate is 95.83 % under the band 890 nm. The fusion result of 3 bands 750, 840, and 980 nm can be improved to 97.13 %. The growth rate is up to 31.18 %.

Table 10.2 Band fusion result

Band number	Optimal band combination (nm)	Recognition rate (%)
1	890	95.83
2	740, 840	96.87
3	750, 840, 980	97.13
4	720, 750, 840, 980	97.27
5	720, 750, 780, 920, 970	97.39

Another experiment is done to testify whether three cluster center bands are enough and effective to represent the whole multispectral region. Recognition rates of more than 3 or less bands are also calculated based on improved hierarchical clustering results in Table 10.2. The comparison result indicates that the fusion of three bands can reach a higher level of recognition performance than two bands fusion. Although the combination of more bands has higher rate, the improvement is not significant. It testifies once again the characterization capability of three bands.

10.4.3 Anti-spoof Test

In addition to the recognition improvement, another distinct advantage of applying multispectral technique on dorsal hand recognition is that it is helpful to liveness detection since hand vein is taken as the extracted feature for representation. Single band recognition usually has the difficulty in distinguishing a real sample and a printed image when the impostor has little difference to the real one on certain band. Multispectral technique can make up single band's deficiency in anti-spoof ability as it has one more data dimension. A non-living sample is usually not able to make the difference to the minimum extend on multibands simultaneously.

The change rule of correlation values between various bands of the same sample can be used as criterion for judging whether the probe sample is a fake one or not. A real dorsal hand should have similar shape as global correlation analysis as shown in Fig. 10.6. Take the three representative bands from final band selection result for example; the distance between 750 and 980 nm should be larger than it from 750 to 840 nm, and the first image row in Fig. 10.14 illustrates the change rule. A printed image with the same dorsal hand is tested as a fake one in our multispectral image capture system. The collected images can pass the single band recognition easily as it have smaller distances to real dorsal hand than tolerable genuine distances, which are 0.764, 0.788, and 0.743 for bands 750, 860, and 990 nm, respectively. However, the distances among representative bands appear to be higher than the ones in first row image, and its change goes against the multi-spectral model. It points out that the reflectance change amplitude of non-living sample does not measure up to a real hand in NIR region.

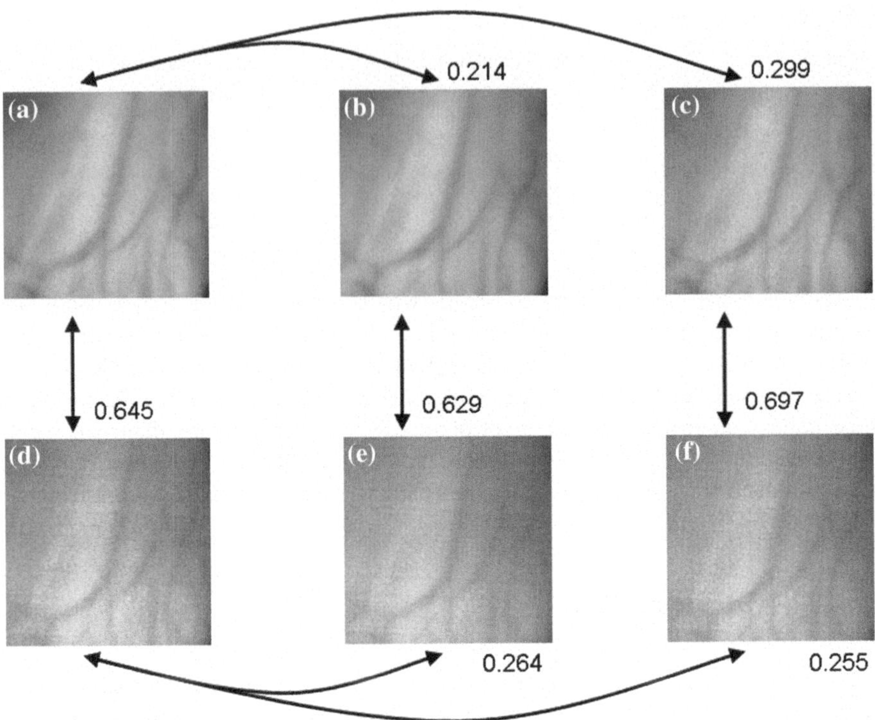

Fig. 10.14 Anti-spoof test based on three representative bands, 750, 840, and 980 nm. **a–c** ROI images of real dorsal hand after preprocessing; **d–f** ROI images of fake one after preprocessing

10.5 Summary

The advantage of multispectral image is to provide complementary information, which is helpful to further recognition rate improvement of dorsal hand. Band clustering can be seen as the preprocessing of band selection. The great benefit is to reduce the subset number and make exhaustive scheme to be feasible in subset generation module. Following the basic rule that neighboring bands should be classified into the same class or divided into neighboring clusters, we use local principle and agglomerative hierarchical clustering method to solve bands' uneven distribution problem and obtain three representative band parts in reaction to the speciality of band distance model. The local principle is not only applied on the initial segmentation, but also reflected in cluster distance normalization. Because the distance matrix is generated from correlations of bands directly, the three representative bands are regarded to own approximate maximum uncorrelation on behalf of the whole band range. The division is testified to be robust enough for different initial conditions in our clustering experiment. With matching-level fusion and additional penalty term, the fusion recognition performance get remarkable promotion.

This work provides guidance to sensor selection in establishment of dorsal hand image capture system. In addition, the improved clustering method is good at dealing with the problem of inhomogeneous data, and it could spread to other research fields such as image segmentation.

References

Arnau TJ, Sotoca JM, Pla F (2004) Non agressive orange acid and sugar indexes estimation system. Image analysis for agricultural products and processes, Vol 69, pp 170–174

Assaleh K, Qaddoumi N, Shanableh T, Adel M (2011) A novel biometric via hand structure using near-field microwave imaging. In: IEEE international conference on automatic face and gesture recognition and workshops, pp 167–172 (doi:10.1109/FG.2011.5771392)

Chen K, Zhang D (2011) Band selection for improvement of dorsal hand recognition. In: IEEE proceedings of international conference on hand-based biometrics, pp 1–4 (doi:10.1109/ICHB.2011.6094333)

Chen S, Zhang R, Su H, Tian J, Xia J (2010) SAR and multispectral image fusion using generalized IHS transform based on a trous wavelet and EMD decompositions. IEEE Sens J 10 (3):737–745. doi:10.1109/JSEN.2009.2038661

Ding Y, Zhuang D, Wang K (2005) A study of hand vein recognition method. In: IEEE proceedings of international conference on mechatronics and automation, vol 4, pp 2106–2110 (doi:10.1109/ICMA.2005.1626888)

Ekenel HK, Stiefelhagen R (2010) Automatic frequency band selection for illumination robust face recognition. In: 20th international conference on pattern recognition, pp 2684–2687 (doi:10.1109/ICPR.2010.658)

Foote J (2000) Automatic audio segmentation using a measure of audio novelty. In: IEEE conference on multimedia and expo, vol 1, pp 452–455 (doi:10.1109/ICME.2000.869637)

Huang R, Li X (2008) Band selection based on evolution algorithm and sequential search for hyperspectral classification. In: Proceeding of international conference on audio, language and image processing, pp 1270–1273

Jia W, Huang D-S, Zhang D (2008) Palmprint verification based on robust line orientation code. Pattern Recogn 41(5):1504–1513. doi:10.1016/j.patcog.2007.10.011

Johnson RA, Wichern DW (1998) Applied multivariate statistical analysis, 4th edn. Prentice-Hall, Upper Saddle River

Kumar A, Prathyusha KV (2009) Personal authentication using hand vein triangulation and knuckle shape. IEEE Trans Image Process 18(9):2127–2136. doi:10.1109/TIP.2009.2023153

Ozawa K (1983) CLASSIC: a hierarchical clustering algorithm based on asymmetric similarities. Pattern Recogn 16(2):201–211. doi:10.1016/0031-3203(83)90023-7

Pearson K (1896) Mathematical contributions to the theory of evolution, regression, heredity and panmixia. Philos Trans R Soc Lond 187:253–318

Sanchit, Ramalho M, Correia P, Soares L (2011) Biometric identification through palm and dorsal hand vein patterns. In: IEEE international conference on computer as a tool (EUROCON), pp 1–4 (doi:10.1109/EUROCON.2011.5929297)

Theodoridis S, Koutroumbas K (2006) Pattern recognition, 3rd edn. Elsevier, Amsterdam

Wang L, Leedham G, Cho S-Y (2007) Infrared imaging of hand vein patterns for biometric purposes. IET Comput Vision 1(3–4):113–122. doi:10.1049/iet-cvi:20070009

Wang K, Zhang Y, Yuan Z, Zhuang D (2008) Hand vein recognition based on multi supplemental features of multi-classifier fusion decision. In: IEEE proceedings of international conference on mechatronics and automation 1790–1795 (doi:10.1109/ICMA.2006.257486)

Wu X, Gao E, Tang Y, Wang K (2010) A novel biometric system based on hand vein. In: 5th international conference on frontier of computer science and technology, pp 522–526 (doi:10.1109/FCST.2010.65)

Yang L, Liu X, Liu Z (2010) A skeleton extracting algorithm for dorsal hand vein pattern. In: International conference on computer application and system modeling 13, pp 92–95 (doi:10.1109/ICCASM.2010.5622671)

Yuksel A, Akarun L, Sankur B (2010) Biometric identification through hand vein patterns. In: International workshop on emerging techniques and challenges for hand-based biometrics, pp 1–6 (doi:10.1109/ETCHB.2010.5559295)

Chapter 11
Comparison of Palm and Dorsal Hand Recognition

Abstract Palm and dorsal hand have received increasing attention during the last two decades. Though the techniques about the two biometrics started later than some other kinds such as face and fingerprint, their properties of uniqueness and permanence have helped them to own unique and stable status in biometric recognition. Strictly analyzing, the two hand-based biometrics features have lots of common characteristics. For example, the way of ROI localization is similar, and the experience can be borrowed from each other. Meanwhile, they both have surface skin texture and deep vein patterns for researching. Nevertheless, recognition performance of palm is much superior to dorsal hand. This phenomenon mainly results from two aspects, the texture difference and spectral difference. The texture difference determines the feature patterns we can extract for recognition, while the spectral difference can take an important role in image qualities. In fact, these two factors are both related to the difference of biologic tissues. Analyzing optical character of palm skin and dorsal hand skin is necessary to explain the distinguishable capability difference. A combined database of palm and dorsal hand is established with 208 volunteers. The particularity is that the database contains palm and dorsal hand images from the same person. The later recognition experiment result is more convincing to testify the superiority of palm, because the palm and dorsal images are all taken in the same time, and we can make sure that the multispectral light source, collecting environment and status of human body, cannot be interference factors in comparison. The result shows that palm has greatly lower equal error rate, no matter under whether optimal single band or optimal multiple bands.

Keywords Biometric difference analysis · Skin model · Combined database · Biological tissue comparison

D. Zhang et al., *Multispectral Biometrics*,
DOI 10.1007/978-3-319-22485-5_11

11.1 Introduction

Driven mainly by increasing requirements in public security, more and more biometric features are brought into identification to against illegal entering. Palm and dorsal hand are two important features, and the related recognition technique has developed into a mature stage. In order to gain better performance, researchers have stated to investigate a variety of fusion, including the fusion of different features, different sensors, and different modalities. Take the palm for examples, and feature of palmprint and palm vein can be presented in one image with proper fusion scheme (Wang et al. 2007). Using multispectral technique, the complementary information under different light source can also be merged together (Zhang et al. 2010). The combination of dorsal vein and palm can realize the fusion on the matching level and gain much lower recognition error rate (Wu et al. 2010).

Feature pattern, light source, and modality are three important components to determine recognition components. The three components are seen as variables, which can be adjusted or chosen to achieve better performance. Of course, here we do not take the influence of learning and testing methods into consideration, and it is not the optimization goal right now in this chapter. The feature pattern is related to not only the type of descriptor as we usually use in feature extraction, but also the type of intuitive object we can use to distinguish one person from another. For instance, palm vein and palmprint are different feature patterns. In practice, we choose one or both of them as the research object. On the other hand, information for identification can be featured with line-based descriptor, texture-based descriptor, or some other sorts. Optimize the feature descriptors for palmprint, and palm vein is necessary for performance optimization. So far, no fusion has ever been involved with the three components simultaneously. By fixing the object of modality, to exploit the optimal feature pattern and light source is the previous task which has been resolved for both palm and dorsal hand. Nevertheless, the rise of fusion performance would meet bottleneck because all the elements for fusion comes from the same modality. No matter how we try to optimize feature pattern and light source, the independency among them is still greatly lower than it among different modalities. That is why some research shows that the fused recognition error rate of palm vein and dorsal vein can decline to nearly zero (Sanchit 2011).

For different modalities, the performance bottleneck levels differ from each other. The highest recognition rate of dorsal hand is 99.17 % under 890 nm based on feature pattern of MFRAT-CompCode, while 790 nm is the optimal light source with minimal EER 0.2220 %. Despite the matching methods, feature pattern and light source play a leading role in performance difference. The feature pattern is determined by the texture of modality itself. Although palm and dorsal hand can both take advantage on dealing with a large amount of data population problem, there still exist difference in the ability to distinguish individuals. By far, no research has ever toughed the area of modality comparison based on pattern diversity and identification performance difference. Heuristic studies started in the domain of physiology and anatomy. Chen classified palmprint into five main patterns based on the three principle line on the palm and further obtained the statistics of the proportion of all these types (Chen et al. 1981). Wu split palm vein into 8 sorts through dissecting and investigating up to 200 dorsal hands (Wu et al. 1985). His division depends on the bifurcation and convergence circumstances of basilica vein and cephalic vein. Correlative achievements can only be seen as a potential theoretical basis for biometric identification, and these pattern analyses are only rough divisions, which is far from enough to meet the needs of distinction between individual. In fact, whether the division result has larger number of patterns cannot really explain the distinction ability of feature pattern. We need more in-depth study about corresponding physiological structure into this question.

On the other side, light source is another factor dramatically affects the final performance. The theoretical basis of optimal band selection is that light source can cause the change in texture. The change does not only include image quality, but also the reflected subcutaneous tissue layer. Generally speaking, the depth light can reach increases, while the wavelength getting larger. Compared with visible light, NIR has the ability of spreading out deeper information before the sensors. Due to the difference in biological tissue composition, the spectral characters of different modalities vary a lot. For example, the proportion of water, adipose, and blood are not constant values in skin at different depth. In order to investigate the spectral influence on collected image, studying more about interaction between skin and light is essential premise. Lots of researchers have ever tried to set up ideal skin model to simulate skin–light interaction process, though the starting point of their work is not relevant with pattern recognition. There is a considerable amount of research on skin model available in biomedical literatures, and it reveals that the related largest application market is disease diagnosis (Tuchin 2000). Understanding how light is absorbed and propagated in skin tissues is the commonality of these research tasks. For example, the lack of melanin in the skin layers of corneum and epidermis would cause vitiligo (Chen et al. 2005). With the help of skin model establishing, people can adjust the model with certain parameters about the melanin to simulate actual patients. In fact, an appropriate skin model is a great benefit to pattern recognition in terms of explaining how the light reflects the useful texture information from different layers. Our work is further using the model to represent spectral difference of palm and dorsal hand. The biophysically based comparison can provide us with reasons explaining why palm performs better than dorsal hand in general.

This chapter also tries to demonstrate this topic based on experimental data. The conclusion that palm is superior to dorsal hand comes from the researches which are focus on palm or dorsal hand only. The biggest problem is the database difference. Specially speaking, it is related to volunteer difference, image acquisition device difference, environmental difference, and sample number difference. Making sure the four factors not to be the influential factors is necessary to comparing work. Following the similar sample acquisition procedure as the previous chapters described, we set up a new combined multispectral database, which has the samples of palm and dorsal hand from the same volunteers. The acquisition of the two modalities is implemented at the same time. Then, the experimental comparing result can be a reliable proof to testify the preliminary conclusion.

11.2 Difference Analysis

The difference between palm and dorsal hand can be divided into three aspects. The most important one is the physiological difference, which determines the texture object. No matter whether the imaging object is palm or dorsal hand, surface texture and subsurface texture can be both reflected on the acquisition picture. However, identification has different priorities on these textures. As it turns out, palm relies more on surface print, while the subsurface vein of dorsal hand plays a more important role in identification. Though the spectral change has the ability to change distribution proportion of surface and subsurface skin, the change is limited. Take dorsal hand for example, surface texture does not have obvious identified special lines, so the performance is not comparable to robust vein pattern. If we adjust the light source to make sure the contribution of surface texture to be bigger than subsurface vein, the total performance would be very low. On the premise of highest recognition rate, the feature pattern is firstly determined. Then, the skin layer the corresponding feature pattern locates constrains the optimal light spectrum in a narrow spectral space. So texture itself is the most important aspect for distinguishing capability level.

We decide to take spectral character difference as the second factor to analyze the two modalities. Now that the optimal feature pattern locates on different skin layers, the case that light absorption and transmission on its path should be studied in detail. Skin model is just the tool for explain why 790 and 890 nm are the optimal bands for palm and dorsal hand, respectively. The last aspect is some other influential factors such as the fine hair, pore, and others.

11.2.1 Physiological Structure Difference

Intuitively, the biggest difference between palm and dorsal hand is the feature pattern. It is worth mentioning that the information richness on palm is obviously bigger than dorsal hand. On one hand, palmprint is surface texture, and reflected light does not need to penetrate into different skin layers. So the reflected light energy is biggest without loss inside the skin, and the texture has shaper gray value change at the places of ridge and valley. Figure 11.1 demonstrates that palmprint often has clearer boundary of the main principle lines. Due to the dissipation of light, small blood vessels cannot be detected easily, and this part may be easily under impact from individual difference, such as the human race and adiposity.

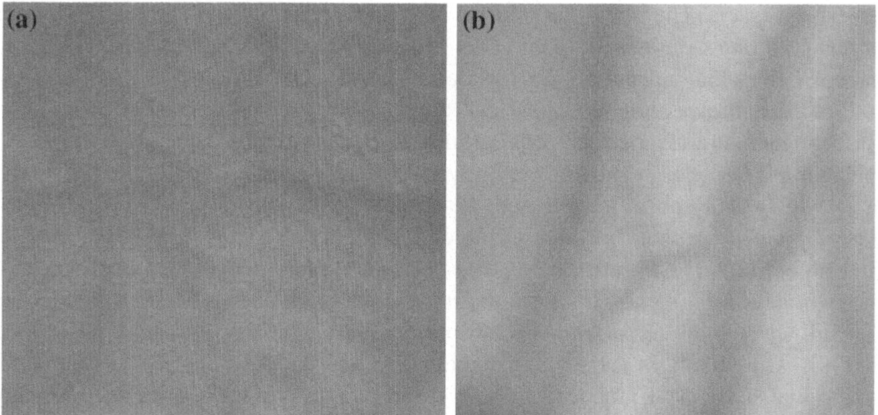

Fig. 11.1 ROI image from the same hand under optimal band **a** palm under 790 nm; **b** dorsal hand under 890 nm

On the other hand, palm vein is also significant for identification. For its optimal band 790 nm, palmprint and palm vein can be captured simultaneously. As we know, the optimal band for palmprint is in visible light region, while vein image can obtain best quality image around 850 nm. Although the imaging effect of palm under 790 may not be able to reach the best point no matter whether for palmprint or palm vein, the integration of the two feature patterns can be adjusted to be optimal one for identifying individuals. When two feature patterns are fused in one image, it demonstrates the great advantage of information complementation.

This section tries to explore the reason of this divergence from physiological structure. The comparison work involves two aspects, surface comparison and subsurface comparison. To the best of our knowledge, palm is one the parts which have the thickest corneum on human body. The thickness can reach as high as 1.6 mm, while the mean value on normal skin such as dorsal hand is usually no more than 0.2 mm (Meglinsky et al. 2001). Corneum is composed mainly of dead cells, embedded in a particular lipid matrix (Talreja et al. 2001). Though the corneum layer is almost transparent, palmprint on such thick layer can still form obvious ridges and valleys depending on reflectance, even if signal-to-noise ratio is low. The existence of palmprint attributes to human requirements for grabbing things. Three principle lines on the palm make clenching fist possible and easily, and other slight wrinkles is benefit to increasing friction force. Obviously, the trend and location of principle lines and wrinkles vary in individuals. Such texture covering whole palm has much more information than subsurface veins, which can only be extracted with large size. The surface of dorsal hand does not have remarkable wrinkles. It only has texture with triangular rid form to adjust the case of skin stretching when we want to clenching fist. Compared with the palm, dorsal hand has better elasticity, and the skin is easily movable because of the rich of superficial fascia. The triangular grid is very thin and small, even the degree change of hand holding would result in great deformation to the grid. So this kind of feature pattern is instable and not proper for identification.

The second aspect of physiological structure comparison concentrates on subsurface, including epidermis and dermis. In the epidermis layer, two types of melanin, phaeomelanin and eumelanin, vary the most greatly in individuals (Thody et al. 1991). This phenomenon is more remarkable among different human race. Parsad pointed out that the value of pigment content can varies from 0.049 to 0.36 (Parsad et al. 2003). Figure 11.2 is the spectral extinction coefficient curves about the two natural pigments. It shows that their absorption ability declines, while the spectral wavelength increasing. The absorption above 700 nm is usually neglected, and it is why NIR light is the better light choice for vein detection. It is essential to notice that palm does not need to worry about such problems because it has no melanin in epidermis layer. The particular character helps visible light also propagate in deep skin tissues as NIR. In this case, the absorptivity of deoxyhemoglobin in vein determines the quality of vessel imaging. Palm vein size is smaller than dorsal hand vein, and it would meet more trouble when we need high accuracy about the feature's location and shape description for identification. Specially speaking, vessels with small width may be extracted to be broken lines, and it would no doubt result in higher matching error.

Good recognition performance of certain modality firstly requires that the feature is robust and distinct enough. With stable principle lines and lots of small wrinkles, palm obtains a thumping win in the surface comparing. Though the palm vein performance is not at par with dorsal hand, the absence of melanin is another advantage to reflect deeper feature patterns clearly. In words, useful information fusion based on multiple feature patterns and no melanin interference make palm recognition to have higher accuracy.

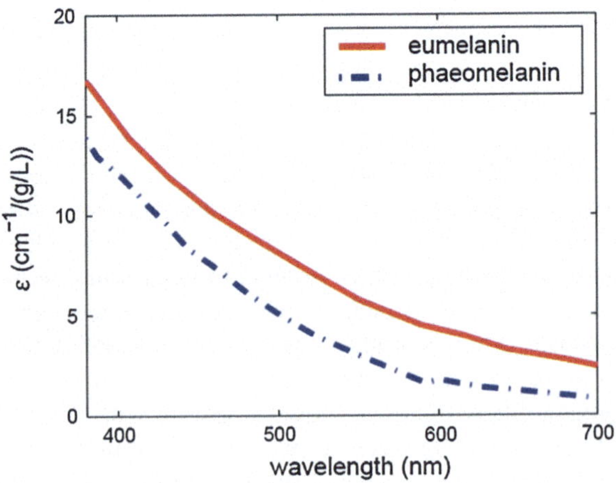

Fig. 11.2 Spectral extinction coefficient curves of eumelanin and pheomelanin in epidermis. Courtesy of S. Prahl and the Oregon Medical Laser Center (OMLC)

11.2.2 Spectral Character Difference

In this section, we want to find biophysically based reason to explain the relationship between optimal bands and the modalities. According to the research results in previous chapter, the best single band for palm and dorsal hand is 790 and 890 nm based on feature optimization. To analyze this, all potential light paths from incidence to emergence should be researched. Because different skin layers have a variety of composition, the skin layers impact on light in different ways. When the light spectrum changes, it would make light transmission more complicated. Establishing a skin model adjusted to spectrum change is a premise to complete corresponding spectral difference analysis.

Due to the flourish of anthropometric since 1980s, people started to know more about skin compositions with the assistance of different sensors. Disease diagnosis becomes a new demand with the technique improvement of anthropometric. For example, measuring bilirubin content in dermis can help to predict erythema condition. Then, researcher started to consider constructing a bridge between symptom and disease and realizing the purpose of rapid diagnosis. The bridge is model setup and computation simulation. With respect to skin-related disease, a skin model is set up with several uncertain parameters for expressing individual difference. For a patient, the relevant parameters are definite by comparing work between simulated model and detail symptom. The case that values out of normal region would be judged as possible disease.

Skin is conventionally divided into 3 layers: epidermis, dermis, and hypodermis. Such structure-based division is not proper for the spectral character analysis. For example, corneum and basal layers which has melanin both belong to epidermis, but the light transmittance vary from each other greatly. So skin spectral model is essential to solve the problem of layer division. Aravind summarizes different kinds of skin spectral models and presents a new Biospec model (Aravind and Gladimir 2004) in his technical report. The schematic cross section of human skin tissues is shown in Fig. 11.3.

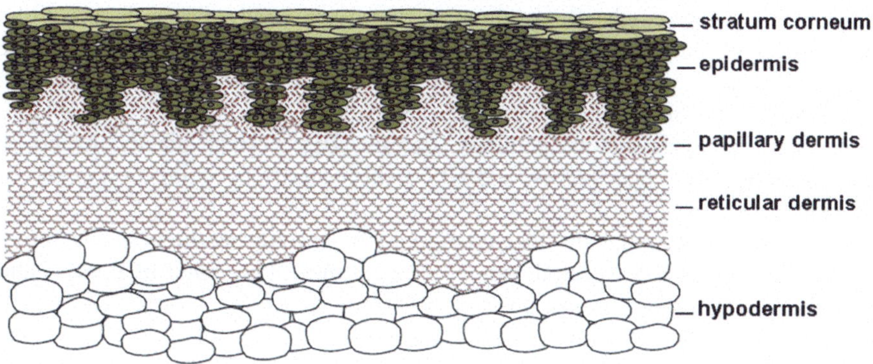

Fig. 11.3 The layer division of BioSpec model (Aravind and Gladimir 2004)

From top to bottom, the layers in BioSpec model are stratum corneum, epidermis, papillary dermis, reticular dermis, and hypodermis, respectively. This division mainly depends on the type of cells. There are two special points worth mentioning about this model. First, different skin layers are seen as different transmission medium. So reflection and refraction occur frequently on medium interface. It means that once the light is reflected to the former interface, part of light could be reflected back and kept in this layer. Second, the author combines the absorption model and scattering model together to realize more accurate simulation. In fact, the absorptivity of water and adipose is limited in visible light region. Different kinds of pigments and blood are the main material absorbing light with spectral region. Most part of light are transmitted in the form of scattering. Surface scattering and subsurface scattering are two aspects influence light transmittance jointly.

It is reported that only 5–7 % of light incident on the stratum corneum is reflected back to the environment (Tuchin 2000). Once the remaining portion of light is transmitted to stratum corneum and epidermis, surface scattering plays the dominant role. The corneum difference between palm and dorsal hand would results in the light transmittance. Surface scattering is relevant with the biological factors such as aging and hydration (Nadia et al. 2002). Aging and dry skin have thinner epidermis layer, and the aspect ratio of cells is close to 0, and light would be diffused slowly. Palm skin on this layer is often rich in water, the cell is more circular, and the reflected light becomes more diffuse. It equals a resistance pressure to the light forward transmission. On the contrary, more light is reflected to the sensor expressing more about the surface texture information.

According to biophysical characters of different layers, subsurface scattering is sorted to be Mie scattering and Rayleigh scattering (Saidi et al. 1995). As the particle in epidermis has similar size with the light wavelength, Mie scattering occurs in this layer. It can be approximately seen as forward scattering. Though the scattering is spectral dependent, the influence is slight and the spectral space just shifts to shorter wavelength a little (Bruls et al. 1984). The biggest difference among the spectral bands is relevant to Rayleigh scattering. Papillary dermis is fully filled with collagen fibers with small scale, which is even smaller than spectral wavelength. Rayleigh scattering is inversely proportional to the forth power of light wavelength. So some part of visible light part would be dispersed greatly and produce backwards scattering. Suffice to say, Rayleigh scattering is the main reason why short wavelength spectral are not able to reflect deeper layers below papillary dermis. Big vessels mainly locate in reticular dermis or deeper places, and NIR light naturally becomes the top choice for vein detection.

It is worth mentioning that deoxyhemoglobin in vein has highest absorption around 780 nm in NIR spectral region. The optimal single band of dorsal hand recognition which relies almost on vein is 890 nm. Since the spectral wavelength would influence the transmitting depth, 780 nm may be not able to reach the skin layer with the richest veins' reticular information. So 890 nm can be seen as trade-off point based on the view of absorption and scattering. For palm identification, without melanin interference on palm, the optimal single band 790 nm is the interaction results of pattern feature, absorption and scattering.

11.2.3 Other Difference

Other difference between palm and dorsal hand discussed in this section can be loosely classified into deterministic difference and non-deterministic difference. Deterministic difference means that the difference exists as a normal form. For example, palm has more sweat glands and pores than dorsal hand. Because contactless-based acquisition system does not need to collect high resolution images, pore is almost invisible, and the difference could not introduce in any influence, whether positive or negative. Fine hair on dorsal hand surface is another deterministic difference; sometimes, this phenomenon is remarkable for males. The felling direction of fine hair is indefinite, and it would result in difference in multiple-sampling for the same person. However, the influence of fine hair can be constrained in visible light, as surface dark texture almost disappears in NIR spectral region. It can be seen that deterministic difference can be neglected in our palm–dorsal hand comparison work.

Non-deterministic difference is related to individual difference while sampling, but the occurrence is also associated with special modality. For example, the probability of palm sweat is much higher than the one of dorsal hand sweat. Palm sweat can reflect incident light with greater energy than other parts. Discrete high light spots may cover useful information and make it fragmented. This is harmful for palm recognition. With the similar way, some skin diseases would also cause bring down the recognition performance of certain modality to a lower level.

According to the spectral model described in previous section, blood's absorption is an important factor influencing the final optimal band and performance. Increasing or reducing blood content in papillary dermis can obviously change the reflected light (Huang et al. 2007). The consequence is connected with both palm and dorsal hand, and it is hard to judge which modality is affected more seriously. Redness of skin change can be caused by squeezing pressure, temperature in acquisition environment, health status, and motion state. Fortunately, all these instable factors can be controlled in sample collecting procedure.

11.3 Comparison Experiment

11.3.1 Combined Database

Regardless of palm or dorsal hand, the former multispectral databases are just proper for the comparing work based on bands. Although we can acquire the conclusion that palm performs better than dorsal hand through rough comparison, it is somewhat hasty to evaluate precisely how much better the palm can be based on these mutually independent databases. Comparison between two modalities should be protected from interference by any factors in sampling process. Strictly speaking, the modality comparison work puts forward higher requirements for database than the conventional research work, such as band comparison and modality fusion. The variables in sampling process include volunteer composition, physical status, and external environment. To eliminate these affects, establishing a combined database to make sure that the sampling object and environment are the same is necessary.

The multispectral images of palm and dorsal hand are both imaged by the filtered light with the wavelength from 520 to 1040 nm. The spectral resolution is 10 nm associated with the filtering character of the system. Apart from the light source with the same distance, two sets of image acquisition device are placed for the fixing and straightening of palm and dorsal hand, respectively. 208 volunteers are all East Asian adults, aging from 20 to 60 years old. Their hands are asked to be recorded twice within half a month, so the final combined database includes 2 sessions with five multispectral groups in each.

According to the mature ROI extraction method, ROI is extracted from original 512×501 size image. By statistical analysis, the ROI sizes of palm and dorsal hand are designed to be 160×160 and 224×224, respectively, with the expectation of covering useful information as more as possible. Then, all ROI images are resized to a uniform value 128×128 for comparing work.

11.3.2 Single Band Comparison

For modality comparison work, whether different modalities should use the uniform feature descriptors is a problem worth considering. The former research reveals the information that optimal band selection is a joint optimization process of band and feature pattern. Separating any factor from the other one would make the relevant result to be unpersuasive. On the other hand, the optimal feature patterns for different modalities are hard to achieve unity. Palm has rich information so the globe texture descriptor is suitable for identification work. The remarkable blood vessels on dorsal hand construct reticular formation, and the line feature extraction is appropriate to extract interference-free information. Hence, we can see that the morphological characteristics can impact on band selection result by feature pattern optimization. Taking any uniform feature pattern as the tool for comparison work would make at least one modality cannot demonstrate its best performance.

The combined database can be separated into two single modality databases. Predictably, the single modality databases can obtain the similar band selection result following the same procedure as introduced in the former chapters, because we ensure that the volunteer coverage is large enough and the acquisition environment does not change significantly. So we decide to take use of the former band selection result directly for palm and dorsal hand comparing work.

Table 11.1 Single band comparison based on palm and dorsal hand

Modality	Feature pattern	Single band (nm)	EER (%)
Palm	CW-SSIM	790	0.44
Dorsal hand	Radon transform	890	1.61

Given the conclusion that the complex wavelet structural similarity distance and Radon transform-based Daugman distance are the optimal feature pattern for palm and dorsal hand, respectively, they are used for different single modality databases. One group of multispectral database in Sect. 11.1 is chosen as gallery samples each time, the samples in Sect. 11.2 are all chosen as probes. Up to 5200 ($5 \times 5 \times 208$) genuine distances and 1,076,400 ($5 \times 5 \times 208 \times 207$) imposter distances are needed for each band in verification test. As the result that the optimal single bands for the two modalities are known, the experiment is just implemented on 790 and 890 nm only. Table 11.1 gives out the equal error rate based on the two single modality database. The distinguishing capability of palm is significantly better than dorsal hand.

11.3.3 Multiple Bands Comparison

No matter palm or dorsal hand, optimal multiple band selection results both resort to maximal uncorrelation principle. Correlation map is used to obtain the final typical bands through proper clustering scheme and exhaustive method. The optimal band combination of palm is 580, 760, and 990 nm, while dorsal hand can peak at the fusion of 750, 840, and 980 nm. Because palmprint and palm vein can both take an active role in performance upgrade, the three optimal multiple bands for palm, respectively, represent palmprint, palm vein, and the band part they works together. By contrast, the drastic reduction of available information in visible light part makes optimal multiples bands have spectral regional distribution, mainly concentrated in NIR light part with good quality imaging. It also indicates that the downside of spectra lower than 750 nm lies in the lack of distinguishable information, while the images under the bands higher than 980 nm would meet trouble of getting more and more blurred because of light scattering with long path.

Due to the fact that dorsal hand relies severely on the hand vein, the span of optimal bands combination is much smaller than the one of palm. It is predicted that performance improvement of multiband dorsal hand fusion would not be as distinct as the palm. To be fair, band fusion of palm and band fusion of dorsal hand both adopt weighted sum score-level fusion. The experiment is designed as the single band comparing work except the additional calculation of fusion distance. Table 11.2 is the fusion results based on two single modality databases.

Table 11.2 Multiple bands comparison based on palm and dorsal hand

Modality	Optimal multiple bands (nm)	EER (%)
Palm	580, 760, 990	0.28
Dorsal hand	750, 840, 980	1.31

The fusion EER of palm can reduce from the single band value 0.44 to 0.28 %. The magnitude of decline is 36.4 % and it attributes to the fusion of multiple feature patterns. Though the palmprint and palm vein are imaged in the same picture and there exists a large amount of redundant information among different bands, the process of optimal band selection can minimal redundancy. Then palm's multiple band fusion is very similar as multiple modality fusion, and the occurrence of performance boost is natural. In contrast, the degree to which the EER of dorsal hand declines from 1.61 to 1.31 % is just 18.6 %. The reason causes the difference can be observed directly from the ROI images in the combined database (Fig. 11.4).

The three principle lines of palmprint are apparent in Fig. 11.4a, while we can hardly find the trace of them under optimal band 990 nm as shown in Fig. 11.4c. The primary differences among the bands in the lower row are some small vessels. Affected by image blurring, the small change is not remarkable unless proper gradient detectors amplify the gray change. To better illustrate the information complementary capability of palm based on these optimal bands, correlation coefficient is calculated to represent the remaining redundancy between the two

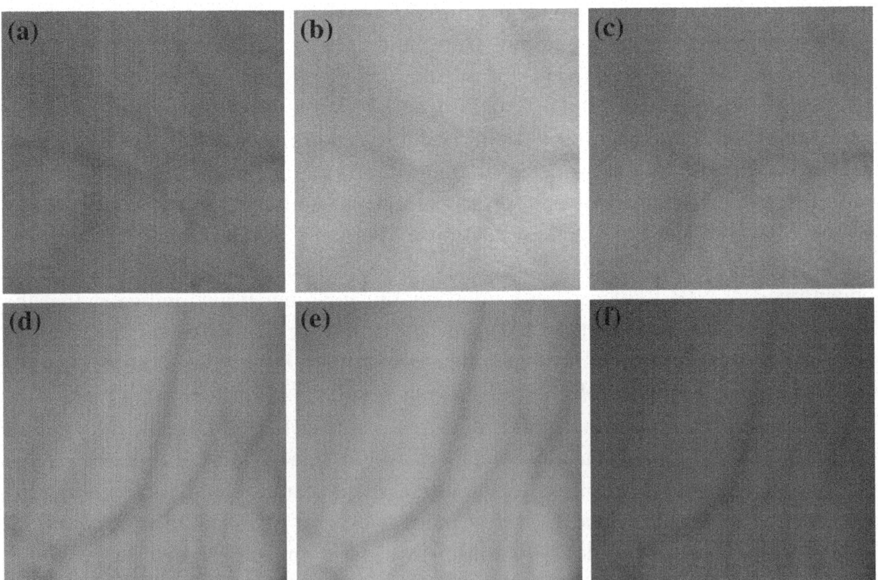

Fig. 11.4 ROI images from the same left hand under optimal bands. The upper row is palm images, and the lower ones are about dorsal hand. **a** 580 nm; **b** 760 nm; **c** 990 nm; **d** 750 nm; **e** 840 nm; **f** 980 nm

ending bands from the optimal band combination. Correlation is little affected by the globe gray change caused by absorption change under different spectra. Considering the consecutive texture change along the spectra wavelength, the two ending bands can cover the largest span the bands combination can express, so the complementary capability between them should be the greatest. Experiments show that the correlation between 580 and 990 nm for palm is only 0.769, while the value of dorsal hand between 750 and 980 nm is 0.966. The lower correlation coefficient of palm is bound to lead higher efficiency for fusion.

11.4 Summary

The optimization of band and feature pattern is performance-oriented result, and it tries to address the issue of best approximation. Nevertheless, different modalities in biometrics vary in the aspect of distinguishing ability. The comparison work in this chapter aims to help us understand better about how feature pattern and spectra interact with each other. First, feature pattern plays a critical role in determining the level of identification capability. A robust feature pattern should own the attribute that it is less affected by instable factors. Secondly, the advantage of spectra controlling is just to adjust the band to the optimal point, which is greatly limited to the layer on which feature pattern locates. So physiological structure difference is the main start point to explain why the performance of palm is superior to dorsal hand. It is found that surface texture corneum is the key point to let palm have the advantages on information richness and image quality. Combined with spectral character difference between palm and dorsal hand, skin spectral model is introduced to explain why the optimal bands are 790 and 890 nm for the two modalities. We conclude that the band selection result is just the trade-off of different components' absorption and light scattering character though skin cells. A palm–dorsal hand combined database is set up to make sure that the two modalities' image are collected at from the person, while the problem of environmental and device difference can also be solved at the same time. The experimental results provide us strong proof that palm can perform significantly better not only under optimal single band, but also under optimal multiband combination.

References

Aravind K, Gladimir V G (2004) A study on skin optics. Technical report, University of Waterlo, Canada. https://www.imt.liu.se/edu/courses/TBMT36/pdf/Text4.pdf. Accessed Jan 2004

Bruls W, Leun J (1984) Forward scattering properties of human epidermal layers. Photochem Photobiol 40(2):231–242. doi:10.1111/j.1751-1097.1984.tb04581.x

Chen R, Huang B, Wang Y, Zeng H, Xie S (2005) Skin optical model. Chin J Acta Laser Biol Sinica 14(6):401–404

Chen Z, Wang Z, Wang P (1981) An analysis of handprint patterns of normal Chinese adults. Chin J Acta Anat Sinica 12(1):61–65

Huang B, Chen R, Zeng H, Wang Y, Xie S (2007) The impact of blood content in skin tissue on skin spectra. J Chin Spectrosc Spectr Anal 27(1):95–98

Meglinsky I, Amtcher S (2001) Modelling the sampling volume for skin blood oxygenation measurements. Med Biol Eng Comput 39(1):44–49. doi:10.1007/BF02345265

Nadia MT, Prem K, Jean LL, Roland B, Dominique B, Bernard Q (2002) A computational skin model: fold and wrinkle formation. IEEE Trans Inform Technol Biomed 6(4):317–323. doi:10.1109/TITB.2002.806097

Parsad D, Wakamatsu K, Kanwar A, Kumar B, Ito S (2003) Eumelanin and phaeomelanin contents of depigmented and repigmented skin in vitiligo patients. Br J Dermatol 149:624–626. doi:10.1046/j.1365-2133.2003.05440.x

Saidi IS, Jacques SL, Tittel FK (1995) Mie and Rayleigh modeling of visible-light scattering in neonatal skin. Appl Optics 34(31):7410–7418

Sanchit, Ramalho M, Correia P, Soares L (2011) Biometric identification through palm and dorsal hand vein patterns. IEEE Int Conf Comput Tool (EUROCON) 1-4. doi:10.1109/EUROCON.2011.5929297

Talreja P, Kasting G, Kleene N, Pickens W, Wang T (2001) Visualization of the lipid barrier and measurement of lipid pathlength in human stratum corneum. AAPS PharmSCi 3(2):1–9. doi:10.1208/ps030213

Thody A, Higgins E, Wakamatsu K, Ito S, Burchill S, Marks J (1991) Pheomelanin as well as eumelanin is present in human dermis. J Invest Dermatol 97(2):340–344

Tuchin V (2000) Tissue optics: light scattering methods and instruments for medical diagnosis. The International Society for Optical Engineering, USA

Wang JG, Yau WY, Suwandy A, Sung E (2007) Fusion of palmprint and palm vein images for person recognition based on "laplacianplam" feature. IEEE Conf Comput Vis Patt Recogn 1-8. doi:10.1109/CVPR.2007.383386

Wu J, Qing Y, Wang Y (1985) The Veins of the Dorsum of the Hand. Chin J Anat 2:014

Wu X, Gao E, Tang Y, Wang K (2010) A novel biometric system based on hand vein. 5th Int Conf Front Comput Sci Technol 522–526. doi:10.1109/FCST.2010.65 2010

Zhang D, Guo Z, Lu G, Zhang L (2010) An online system of multispectral palmprint verification. IEEE Trans Instrum Meas 59(2):480–490. doi:10.1109/TIM.2009.2028772

Part V
Conclusion and Future Work

Chapter 12
Book Review and Future Work

Biometric recognition, the use of the human physiological and behavioral characteristics for personal authentication, has a long history. In fact, we use it every day. We commonly recognize people based on their face, voice, and gait. Signatures are recognized as an official verification method in legal and commercial transactions. Fingerprints and DNA have been considered effective methods for forensic applications including investigating crimes, identifying bodies, and determining parenthood. Recently, more and more effort has been put on developing effective multispectral biometric systems for various security demands.

This book unveils automatic techniques for multispectral biometrics authentication, from multispectral data imaging, feature extraction, and matching to multispectral biometrics system development. It may serve as a handbook of multispectral biometric authentication and be of use to researchers and students who wish to understand, participate, and/or develop a multispectral biometric authentication system. It would also be useful as a reference book for a graduate course on biometrics.

In this chapter, we first recapitulate the contents of this book in Sect. 12.1. Then, Sect. 12.2 discusses the future of multispectral biometrics research.

12.1 Book Recapitulation

This book has five parts divided into twelve chapters which discuss the multispectral biometrics technology ranging from the hardware design of a multispectral biometrics data acquisition to the algorithm designed for multispectral biometrics preprocessing, feature extraction, and matching.

Chapter 1 introduces recent developments in biometrics technologies, some key concepts in biometrics, and the importance of developing new biometrics: multispectral biometrics.

Chapter 2 gives an overall review of current multispectral imaging technologies and their applications in biometrics, including multispectral face, multispectral iris, multispectral palm, multispectral fingerprint, and multispectral dorsal hand.

D. Zhang et al., *Multispectral Biometrics*,
DOI 10.1007/978-3-319-22485-5_12

Chapter 3 presents a novel multispectral iris acquisition system. The design and implementation of a high-speed multispectral iris capture device is described. A multispectral iris database is created, and iris image-level fusion is studied to investigate the effectiveness of the proposed capture device by the 1-D Log-Gabor wavelet filter.

Chapter 4 uses East Asian irides as research subjects and explores the possibility of clustering wavelengths based on the maximum dissimilarity of iris textures. It is empirically found that 3 clusters are enough to represent the 10 feature bands of spectral wavelengths from 545 to 940 nm, using the agglomerative clustering based on an improved multigroup two-dimensional principal component analysis ($(2D)^2$PCA).

Chapter 5 shows a prototype of multispectral iris recognition system. The proposed dual-eye capture device benefits from simplicity and lower cost. It is designed to have good performance at a reasonable price for civilian personal identification applications. It is the first attempt to design the dual-eye multispectral iris capture handheld system based on single low-resolution camera.

Chapter 6 presents an online multispectral palmprint authentication system. A data acquisition device is designed to capture the palmprint images under blue, green, red, and NIR illumination in less than 1 s. Our experimental results show that the red channel achieves the best result, while the blue and green channels have comparable performance but are slightly inferior to the NIR channel. A new score-level fusion scheme is proposed to integrate the multispectral information.

Chapter 7 analyzes the palmprint recognition performance under seven different illuminations to investigate whether the most widely used white light is the best illumination for palmprint recognition. Experimental results show that white light is not the optimal illumination, while yellow or magenta light could achieve higher palmprint recognition accuracy than the white light.

Chapter 8 presents a study on feature band selection for multispectral palmprint. By analyzing hyperspectral palmprint data (520–1050 nm), our experimental results show that 3 spectral bands could provide most of the discriminate information of palmprint. This finding could be used as the guideline for designing new online multispectral palmprint systems.

Chapter 9 studies the optimal spectral for dorsal hand recognition. We collected a multispectral dorsal hand database including images by visible light and NIR. By investigating different feature extraction methods, it is empirically found that 890 nm is the best choice for dorsal hand recognition.

Chapter 10 tries to address two basic issues of multispectral dorsal hand, the optimal number of feature bands and which bands can express multispectral model more precisely for dorsal hand. Local principle and agglomerative hierarchical clustering method are used to solve uneven distribution problem. We empirically find that 3 representative bands could convey most of the useful information for dorsal hand.

Chapter 11 investigates two hand-based biometrics traits, palm hand and dorsal hand. These two traits are both related to the difference of biologic tissues. Analyzing optical character of palm skin and dorsal hand skin is necessary to explain the distinguishable capability difference. So we build a new multimodal biometric data including these traits. Experimental results show that palm has greatly lower error rates than dorsal hand, no matter using single optimal feature band or multiple optimal bands.

12.2 Future Work

We have developed multispectral iris (Part II), multispectral palmprint (Part III), and multispectral dorsal hand (Part IV) recognition prototypes already. Each system has its own characteristics and can handle different real-world challenges accordingly. Although these prototypes are successful, there are still some research aspects that need for further study:

(a) Sensor size
(b) Higher performance
(c) Distinctiveness
(d) Permanence
(e) Privacy concerns

12.2.1 Sensor Size and Cost

Now the sizes of multispectral imaging biometrics sensors are larger than traditional sensors. The same situation lies in their costs. These two factors will greatly affect the application of multispectral biometrics technologies. For example, traditional fingerprint sensors are very small now, and it can be used in many fields, such as mobile phone, notebook, and electronic lock, but the size of multispectral fingerprint sensor is a little big to these applications. So there are lots of work should be done in the future to control the sensor sizes and costs.

12.2.2 Higher Performance

Higher performance is the main object for all biometrics systems, even though multispectral biometrics technologies' performance is better than corresponding single spectral technologies, and there are still spaces to be improved for large-scale population, such as national citizen ID. More research work can be done on feature extraction and matching. Another research direction is to fuse multispectral features efficiently, such as the fusion scheme in Chap. 6.

12.2.3 Distinctiveness

It is not uncommon to ask a question about a biometrics, whether it is unique enough for a relatively large user database. For the distinctiveness, here we interest in the information in different multispectral biometrics, whether they are sufficiently enough for identifying a person from large population. In other words, can we find out some multispectral biometric candidates from different persons, but they are very similar? To investigate the distinctiveness of multispectral biometrics, a large-scale database is necessary; until now, our biggest multispectral palmprint database only includes 6000 samples from 500 palms; and the multispectral iris database just contains 4800 images from 40 volunteers. We should pay more attention to the database construction in the later work and then build a theoretical model to analyze their distinctiveness.

12.2.4 Permanence

Permanence is another important issue for biometric identification. Each biometrics has some variations. Even for DNA, mutation is one of the means to change it. Our weight, age, and living styles may change our palmprints, and irides have similar situations. But these biometrics characteristics cannot frequently change, some features may be permanent, and some features are stable in a period of time. Our previous study shows that our method can recognize multispectral palmprints collected within a period of time. We need more data to analyze their permanence.

12.2.5 Privacy Concerns

Privacy concern of a biometrics identifier is another important issue affecting the deployment of a particular biometrics. Like fingerprint technology, some people may be afraid of their fingerprint data being used by third parties for criminal investigation leading them reluctant to adopt it. In order to solve this problem, different aspects should be taken care of. In the future, we can consider adopting some encryption methods on the multispectral biometric data of each of the communication channels involved so that no one can take the templates out of the system for the enrollment on other applications.

Index

D. Zhang et al., *Multispectral Biometrics*,
DOI 10.1007/978-3-319-22485-5

The manufacturer's authorised representative in the EU is Springer Nature Customer Service Centre GmbH, Europaplatz 3, 69115 Heidelberg, Germany. If you have any concerns regarding our products, please contact ProductSafety@springernature.com

Printed and bound by CPI Group (UK) Ltd, Croydon, CR0 4YY
15/07/2026
02167620-0001